A Student Handbook for Writing in Biology

Fifth Edition

A STUDENT HANDBOOK FOR

Writing in Biology

FIFTH EDITION

Karin Knisely
Bucknell University

Cover photographs by Andrew Sinauer.

A Student Handbook for Writing in Biology, Fifth Edition

Sinauer Associates is an imprint of Oxford University Press.

Address editorial correspondence to:
Oxford University Press, 23 Plumtree Road, Sunderland, MA 01375 U.S.A.

Address orders to:
MPS/W. H. Freeman & Co., Order Department, 16365 James Madison Highway, U.S. Route 15, Gordonsville, VA 22942 U.S.A.

Library of Congress Cataloging-in-Publication Data

Names: Knisely, Karin.
Title: A student handbook for writing in biology / Karin Knisely, Bucknell University.
Other titles: Writing in biology
Description: Fifth edition. | Sunderland, Massachusetts : Sinauer Associates, Inc., [2017] | Includes bibliographical references and index.
Identifiers: LCCN 2017001834 | ISBN 9781319121815 (pbk.)
Subjects: LCSH: Biology--Authorship--Style manuals.
Classification: LCC QH304 .K59 2017 | DDC 808.06/657--dc23
LC record available at https://lccn.loc.gov/2017001834

Printed in U.S.A.
6 5 4 3

To those who go the extra mile in pursuit of excellence

Contents

Preface XIII

CHAPTER 1 ***The Scientific Method 1***

An Introduction to the Scientific Method 1

Ask a question 2
Look for answers to your question 2
Turn your question into a hypothesis 2
Design an experiment to test your hypothesis 3
Record data 5
Summarize numerical data 6
Analyze the data 7
Try to explain the results 7
Revise original hypotheses to take new findings into account 8
Share findings with other scientists 8

CHAPTER 2 ***Developing a Literature Search Strategy 9***

Databases and Search Engines for Scientific Information 12

Comparison of databases 13

Database Search Strategies 15

Understand your topic 15
Define your research goals 16
Subdivide your topic into concepts 16
Choose effective keywords 17
Connect keywords with the operators and, or, *or* not *18*
Use truncation symbols for multiple word endings 19
Search exact phrase 19
Use the same keywords in a different database or search engine 19

Evaluating Search Results 19
Finding related articles 23
Obtaining full-text articles 23
Managing References (Citations) 24
ProQuest RefWorks 24

CHAPTER 3 Reading and Writing Scientific Papers 31

Types of Scientific Communications 31
Hallmarks of Scientific Writing 32
Scientific Paper Format 32
Styles for Documenting References 34
Strategies for Reading Journal Articles 35
Acquire background information on the topic 36
Formulate questions 36
Read selectively 38
Recite 38
Review 38
Share your knowledge 39
Strategies for Reading Your Textbook 39
Survey the content 39
Go to class and take notes 40
Formulate questions 40
Read selectively 40
Recite 41
Review 41
Concept (mind) mapping 41
How to Succeed in College 42
Study Groups 43
Plagiarism 44
Information that does not have to be acknowledged 44
Information that has to be acknowledged 44
Paraphrasing the source text 45
The Benefits of Learning to Write Scientific Papers 46
Credibility and Reputation 47
Model Papers 47

CHAPTER 4 Step-by-Step Instructions for Preparing a Laboratory Report or Scientific Paper 49

Timetable 49
Format your report correctly 51
Computer savvy 51

Getting Started 53
Reread the laboratory exercise 53
Organization 53
Audience 53
Writing style 54
Start with the Materials and Methods Section 55
Tense 55
Voice 55
Level of detail 55
Do the Results Section Next 59
Preparing visuals 60
Tables 62
Line graphs (XY graphs) and scatterplots 64
Bar graphs 70
Pie graphs 72
Organizing your data 73
Writing the body of the Results section 75
Equations 80
Make Connections 81
Write the discussion 81
Write the introduction 84
Effective Advertising 87
Write the abstract 87
Write the title 88
Documenting Sources 89
The Name-Year system 90
The Citation-Sequence system 92
The Citation-Name system 94
Unpublished laboratory exercise 94
Personal communication 95
Internet sources 95
Journal articles 96
Databases 98
Homepages 99
Emails and discussion lists 101

CHAPTER 5 Revision 103

Getting Ready to Revise 103
Take a break 103
Slow down and concentrate 104
Think of your audience 104

Editing 104

Evaluate the overall structure 104
Do the math at least twice 105
Organize each section 106
Make coherent paragraphs 106
Write meaningful sentences 108
Choose your words carefully 112
Construct memorable visuals 120

Proofreading: The Home Stretch 121

Make subjects and verbs agree 121
Write in complete sentences 122
Revise run-on sentences 122
Spelling 124
Punctuation 125
Abbreviations 129
Numbers 129
Format 132

Get Feedback 133

Tips for being a good peer reviewer 134
Have an informal discussion with your peer reviewer 135
Feedback from your instructor 135

Biology Lab Report Checklist 138

CHAPTER 6 *Sample Student Laboratory Reports 141*

A "Good" Sample Student Lab Report 141
Laboratory Report Errors 149
A Lab Report in Need of Revision 154

CHAPTER 7 *Poster Presentations 163*

Why Posters? 163
Poster Format 164

Layout 164
Appearance 164
Font (type size and appearance) 165
Nuts and bolts 166

Making a Poster in Microsoft PowerPoint 166

Design 166
Adding text, images, and graphs 166
Aligning objects 168
Proofread your work 168
Final printing 169

Poster Content 169
Title banner 169
Introduction 170
Materials and methods 170
Results 170
Discussion or Conclusions 171
Literature citations 171
Acknowledgments 171
Presenting Your Poster 172
Evaluation Form for Poster Presentations 172
Sample Posters 172

CHAPTER 8 *Oral Presentations 173*

Structure 173
Plan Ahead 174
Prepare the First Draft 175
Make the Slides Audience-Friendly 175
Focus on one idea at a time 176
Write complete thoughts 176
Use more visuals and fewer words 177
Keep graphs simple 177
Make text easy to read 179
Provide ample white space 179
Use animation feature to build slide content 180
Appeal to your listeners' multiple senses 180
Deal proactively with lapses in audience attention 180
Rehearsal 181
Delivery 182
Presentation style 182
Interacting with the audience 183
Group presentations 184
Fielding listener questions 184
Slide Decks as References 184
Feedback 185

APPENDIX 1 *Word Processing in Microsoft Word 189*

Introduction 189
Good Housekeeping 190
Organizing your files in folders 190
Accessing files and folders quickly 191
Naming your files 192

Saving your documents 192
Backing up your files 193
AutoCorrect 193
Long words 194
Expressions with sub- or superscripts 196
Italicize scientific names of organisms automatically 196
Endnotes in Citation-Sequence System 197
Equations 198
Feedback Using Comments and Track Changes 199
Format Painter 201
Formatting Documents 201
Margins 201
Paragraphs 202
Page numbers 203
Proofreading Your Documents 204
Spelling and grammar 204
Prevent section headings from separating from their body 204
Prevent figures and tables from separating from their caption 205
Errant blank pages 205
Finally, print a hard copy 205
Sub- and superscripts 205
Symbols 206
Tables 208

APPENDIX 2 Making Graphs in Microsoft Excel 213
Introduction 213
Handling computer files 214
Formulas in Excel 214
Writing formulas 215
Copying formulas using the fill handle 220
Copying cell values, but not the formula 220
Formatting the Spreadsheet 220
Wrap text 220
Increase or decrease decimal 220
Format cells 221
Sort data 221
Tables 221
The Page Layout tab 223
Views 223
Adding worksheets to the same file 224
Plotting XY Graphs (Scatter Charts) 224

Saving and Applying Chart Templates 236
Adding Data after Graph Has Been Formatted 237
To incorporate additional data points in the same series 237
To incorporate additional lines on the same graph 237
To change the legend titles 238
Multiple Lines on an XY Graph 239
Plotting Bar Graphs 239
Clustered Column Charts 245
Pie Graphs 246
Error Bars and Variability 249
Adding error bars about the means 251
Data analysis with error bars 253

APPENDIX 3 Preparing Oral Presentations with Microsoft PowerPoint 255

Introduction 255
Handling computer files 257
Planning Your Presentation 258
Designing Slides 258
Selecting a theme 258
Choosing a slide size 260
Views 260
Slide layouts 260
Reusing slides from other presentations 261
Adding animation 262
Importing Excel graphs 263
Working with shapes 263
Adding videos 265
Text formatting shortcuts 266
Delivering Presentations 267
Presenter View 267
Rehearsal 267
Resources at the presentation site 269
Presenter tools 270
Speaker notes, handouts, and sharing electronic presentations 272

Bibliography 273

Index 277

Preface

The goal of this handbook is to provide students with a practical, readable resource for communicating their scientific knowledge according to the conventions in biology. Chapter 1 introduces the scientific method and experimental design. Because the scientific method relies heavily on the work of other scientists, students must learn how to locate primary references efficiently, quite often using article databases and scholarly search engines (Chapter 2). Once a suitable journal article has been found, the next challenge is to read and understand the content. Chapter 3 describes scientific paper tone and format, provides strategies for reading technical material, emphasizes the importance of paraphrasing when taking notes, and gives examples of how to present and cite information to avoid plagiarism. Using the standards of journal publication as a model, students are then given specific instructions for writing their own laboratory reports with accepted format and content, self-evaluating drafts, and using peer and instructor feedback to refine their writing (Chapters 4–6). Besides writing about it, scientists communicate scientific knowledge through posters and oral presentations. How these presentation forms differ from papers in terms of purpose, content, and delivery is the subject of Chapters 7 and 8.

Scientific communication requires more than excellent writing skills—it requires technical competence on the computer. Most first-year students have had little experience producing Greek letters and mathematical symbols, sub- and superscripted characters, graphs, tables, and equations. Yet these are characteristics of scientific papers that require a familiarity with the computer beyond basic keyboarding skills. Furthermore, most first-year students are used to doing calculations on a handheld calculator. When they learn how to apply Excel's formulas to carry out repetitive calculations, their time spent on data analysis decreases markedly. In addition, students need to know how to plot the reduced data to help them understand their results and to format their graphs in a manner that is familiar to other scientists. Finally, good presentation skills require not only good public speaking skills but also an understanding of the kinds of visual aids that make a talk engaging for the audience. For all of these reasons, about one-third of the book is devoted to Microsoft Word, Excel, and PowerPoint features that enable scientists to produce professional quality papers, graphs, posters, and oral presentations effectively and efficiently.

In the Fifth Edition, the appendices have been updated for Microsoft Office 2013 and Office 2016 for Mac. Video tutorials for both Mac and Windows, available from <http://sites.sinauer.com/Knisely5E>, replace many of the screen shots in the previous editions. The videos not only make the book shorter, they give students the visual resources that many of them prefer. Specifically, the tutorials provide time-saving tips for formatting document elements in Word and applying formulas, making graphs, and saving graphs as chart templates in Excel.

In addition to updating and streamlining the appendices, I have reworked almost all of the chapters of the Fifth Edition. Chapter 2 describes how to use scholarly databases in conjunction with Google Scholar (whose scholarly value has increased over the past three years) to find and acquire relevant full-text journal articles. Instructions for the newest version of ProQuest RefWorks are given to illustrate how reference management software facilitates in-text and end reference formatting. In Chapter 3, the sections on reading journal articles and textbooks have been expanded to reflect best practices recommended by university teaching and learning centers. In particular, strategies are provided to help students improve comprehension and retention of what they read. When applied effectively, these same strategies also help students avoid plagiarism.

Chapter 4 provides step-by-step instructions for writing lab reports. The prominent headings throughout the Fifth Edition make it easier to find the important concepts and highlight the specific kinds of problems encountered by students who have little experience writing scientific papers in biology. Many examples of faulty writing and how to correct it are given. The section on writing the introduction has been expanded to help students decide how much and what kind of background information is needed. Similarly, the section on writing the discussion is augmented with an analysis of a published journal article to illustrate how the authors build a strong case for their conclusions. Documentation style has been updated in accordance with the latest edition of the Council of Science Editors Manual (2014). Chapter 5 describes a systematic approach to revision. A new section on instructor feedback has been added, which illustrates how grading electronic versions of student lab reports can save time and improve consistency. For instructors who prefer to grade assignments on paper, Chapter 6 now has a short list of proofreading marks as well as abbreviations and explanations for comments on the kinds of errors typically made by beginning writers. These tables are available on the website as a Word file so that instructors can customize the comments for specific assignments. Students will also find these resources useful to avoid making these errors in the first place. On the flip side, to show students what to include in a well-written lab report, there are numerous short checklists throughout the book, and the Biology Lab Report Checklist has been expanded. In response to user feedback, a lab report in need of revision has been added to Chapter 6. The annotations in the margin of this and the "good" student lab report illustrate characteristics of scientific writing that pertain to both content and style.

Chapter 8 on preparing oral presentations has been updated to reflect the fact that PowerPoint slide decks often serve multiple, not necessarily compatible, functions. These functions include visual aids for the audience, prompts for the speaker, and reference material for those who cannot attend the talk. Some options for structuring slide decks to meet these different needs are described. In addition, Chapter 8 has a new section on dealing with lapses in audience attention during a talk.

While some users of this book may enjoy reading it cover to cover, the majority will use it primarily as a look-up reference. Most of the sections are designed to stand alone so that readers can look up a topic in the index and find the answer to their question. Those who want to learn more about the topic have the option of reading related sections or entire chapters. The Fifth Edition is also available in an e-version for those who prefer to "pack light" and search rapidly and efficiently.

The book is augmented by ancillary materials available on the Sinauer Associates Web page. A biology lab report template in Microsoft Word provides prompts that help students get used to scientific paper format and content. The Biology Lab Report Checklist can be printed out to help students self-evaluate or peer review lab reports. Instructors and students will find the list of proofreading marks and laboratory report comments handy for use in both the revision and feedback stages. The Evaluation Form for Oral Presentations enables listeners to provide feedback to the speaker on things that he/she is doing well as well as areas that need improvement. Similarly, the Evaluation Form for Poster Presentations can be used as a checklist for the presenter and an evaluation tool for visitors during the actual poster session. To illustrate principles of designing effective posters, sample posters are posted online, and each poster is accompanied by a short evaluation of the layout and content. All of the above documents can be downloaded from <http://sites.sinauer.com/Knisely5E>.

Acknowledgments

Many of the improvements in the Fifth Edition were made in response to feedback from users of the Fourth Edition of this book. In particular, I would like to thank Angela Currie (Niagara University), Sharon Hyak (Victoria College), Javier Izquierdo (Hofstra University), Ross E. Koning (Eastern Connecticut State University), Brenda Leicht (University of Iowa), Keenan M.L. Mack (Illinois College), Katharine Northcutt (Mercer University), Luciana Cursino Parent (Hobart and William Smith Colleges), Donald A. Stratton (University of Vermont), and Amy Wiles (Mercer University) for their suggestions. I also want to thank my students in Introduction to Molecules and Cells (BIOL205) and Organismal Biology (BIOL206) laboratories for helping me understand the difficulties they encounter when writing lab reports. My colleagues in the Biology Department at Bucknell University continually provide me with fresh and refreshing ideas on teaching, and I am proud to be part of such a collegial group. I am indebted to the helpful professionals in Library and Information Technology at Bucknell for answering my research and computer-related questions, especially Debra Balducci (technology tutorials and photographing posters), Brianna Derr (production of the video tutorials), Kathleen McQuiston (article databases and scholarly search engines), and Jim Van Fleet (RefWorks and scholarly research). I would also like to thank Katrina Knisely, Sandy Field, and Kathy Shellenberger for providing feedback on various chapters and for many enjoyable conversations. Finally, I would like to thank my children, Katrina, Carleton, and Brian, for their insight into what motivates today's college students and for keeping me up-to-date on the latest technological developments. Brian especially has always been willing to share his technical knowledge with me, communicating concepts with empathy and clarity.

I also express my sincere appreciation to those who contributed to the previous editions: John Woolsey (poster presentations), Lynne Waldman (good student lab report), Marvin H. O'Neal III (Stony Brook University), Joe Parsons (University of Victoria), Randy Wayne (Cornell University), John Byram (W.W. Norton & Co.), Kristy Sprott (The Format Group LLC), Wendy Sera (Baylor University), Elizabeth Bowdan (University of Massachusetts at Amherst), Mary Been (Clovis Community College), Walter H. Piper (Chapman University), Warren Abrahamson (Bucknell University), Karen Patrias (National Library of Medicine), and the staff of the Bucknell Writing Center.

I am grateful to all the professionals at Sinauer Associates and Macmillan Learning who helped with the production of this book, especially Andy Sinauer, who provided encouragement, feedback, and organizational input on all five editions. I would also like to thank Dean Scudder for starting the conversation on the new editions and for help with various requests along the way. Rachel Meyers, Carol Wigg, Alison Hornbeck, and Christopher Small guided the Fifth Edition through the editorial and production processes. Jefferson Johnson deserves the credit for the cool cover, and Jason Dirks and Nate Nolet set up the website for the ancillary materials.

Finally, I would like to thank my parents as well as my husband, Chuck, for their steadfast love and support. They have always encouraged me to do my best, to seek knowledge, and to share it with others. I am grateful to them for giving me the all-important time and space to pursue my goals and passions.

KARIN KNISELY • LEWISBURG, PA

January 2017

Chapter 1

The Scientific Method

Trying to understand natural phenomena is human nature. We are curious about why things happen the way they do, and we expect to be able to understand these events through careful observation and measurement. This process is known as the scientific method, and it is the foundation of all knowledge in the biological sciences.

An Introduction to the Scientific Method

The scientific method involves a number of steps:

- Asking questions
- Looking for sources that might help answer the questions
- Developing possible explanations (hypotheses)
- Designing an experiment to test a hypothesis
- Predicting what the outcome of an experiment will be if the hypothesis is correct
- Collecting data
- Analyzing data
- Developing possible explanations for the experimental results
- Revising original hypotheses to take into account new findings
- Designing new experiments to test the new hypotheses (or other experiments to provide further support for old hypotheses)
- Sharing findings with other scientists

Most scientists do not rigidly adhere to this sequence of steps, but it provides a useful starting point for how to conduct a scientific investigation.

Ask a question

As a biology student, you are probably naturally curious about your environment. You wonder about the hows and whys of things you observe. To apply the scientific method to your questions, however, the phenomena of interest must be sufficiently well defined. The parameters that describe the phenomena must be measurable and controllable. For example, let's say that you learned that:

> Dwarf pea plants contain a lower concentration of the hormone gibberellic acid than wild-type pea plants of normal height.

You might ask the question:

> Does gibberellic acid regulate plant height?

This is a question that can be answered using the scientific method, because the parameters can be controlled and measured. On the other hand, the following question could not be answered easily with the scientific method:

> Will the addition of gibberellic acid increase a plant's sense of well-being?

In this example, "a sense of well-being" is not something that can be measured or controlled.

Look for answers to your question

There is a good chance that other people have already asked the same question. That means that there is a good chance that you may be able to find the answer to your question, if you know where to look. Secondary references such as your textbook, encyclopedias, and information posted on the websites of university research groups, professional societies, museums, and government agencies are usually easier to comprehend than journal articles and may be good places to begin finding answers (see the section "Understand your topic" in Chapter 2). Curiously, attempts to answer the original question often result in new questions, and unexpected findings lead to new directions in research. By reading other people's work, you may think of a more interesting question, define your question more clearly, or modify your question in some other way.

Turn your question into a hypothesis

As a result of your literature search or conversations with experts, you may now have a tentative answer to your original (or modified) question.

Now it is time to develop a hypothesis. A **hypothesis** is a possible explanation for something you have observed. **You must have information before you can propose a hypothesis!** Without information, your hypothesis is nothing more than an uneducated guess. That is why you must look for possible answers before you can turn your question into a hypothesis.

A useful hypothesis is one that can be tested and either supported or negated. A hypothesis can never be *proven* right, but the evidence gained from your observations and/or measurements can *provide support for* the hypothesis. Thus, when scientists write papers, they never say, "The results prove that..." Instead, they write, "The results suggest that..." or "The results provide support for..."

You might transform your question "Does gibberellic acid regulate plant height?" into the following testable hypothesis:

Good: The addition of gibberellic acid to dwarf plants will allow them to grow to the height of normal, wild-type plants.

This hypothesis provides specific expectations that can be tested. In contrast, the following hypothesis is not specific enough:

Vague: The addition of gibberellic acid will affect the height of dwarf plants.

Design an experiment to test your hypothesis

In an **observational study**, scientists observe individuals and measure variables of interest without trying to control the variables or influence the response. While observations provide important information about a group, it is difficult to draw conclusions about cause and effect relationships because multiple factors affect the response. That's the main reason why scientists conduct experiments. **Experiments** are studies in which the investigator imposes a specific treatment on a person or thing while controlling the other factors that might influence the response.

The first step in designing an experiment is to determine which variables might be influential. Of those variables, only one may be manipulated in any given experiment; the others have to remain constant. The individuals in the experiment are then divided into treatment and control groups. The treatment group is subjected to the independent variable and the control group is not; all other conditions are the same for the two groups. If the hypothesis is supported, the individuals in the treatment group will respond differently from those in the control group. If there is no difference in response between the treatment and control groups, the so-called **null hypothesis** is supported. Having enough replicates lends assurance that the results are reliable.

Define the variables Variables are commonly classified as independent or explanatory variables, dependent or response variables, and controlled variables. The *one* variable that a scientist manipulates in a given experiment is called the **independent variable** or the explanatory variable, so called because it "explains" or influences the response. It is important to manipulate *only one* variable at a time to determine whether or not a cause and effect relationship exists between that variable and an individual's response. The other variables that may affect the response must be carefully controlled so that they do not confound the relationship between the independent variable and the dependent variables.

Dependent variables are those affected by the imposed treatment; in other words, they represent an individual's response to the independent variable. Dependent variables are variables such as size, number of seeds produced, and velocity of an enzymatic reaction, which can be measured or observed.

The hypothesis proposed earlier involves testing whether there is a cause and effect relationship between gibberellic acid (GA) treatment and plant height. GA level is the variable that will be manipulated; plant height is the response that we'll measure. Because plant height is affected by many other factors such as ambient temperature, humidity, age of the plants, day length, amount of fertilizer, and watering regime, however, we must keep these controlled variables constant so that any differences in response can be attributed to the GA treatment.

Set up the treatment and control groups The individuals in the experiment are assigned randomly to either a treatment group or a control group. Those in the treatment group will be subjected to the independent variable (GA in this case), while those in the control group will not. Depending on the hypothesis, the control group may be subdivided into positive and negative controls. Negative controls are not treated with the independent variable and are not expected to show a response. Positive controls represent a reference for treatment groups that demonstrate a response consistent with the hypothesis.

HYPOTHESIS: Adding GA to dwarf plants will allow them to grow to the height of normal, wild-type plants.

TREATMENT GROUP: Dwarf plants + GA

CONTROL GROUPS:

NEGATIVE: Dwarf plants + no GA (substitute an equal volume of water)

POSITIVE: Wild-type plants + no GA

Determine the level of treatment for the independent variable How much GA should be added to the dwarf plants in the treatment group to produce an increase in height? Too little GA may not effect a response, but too much might be toxic. To determine the appropriate level of treatment, consult the literature or carry out a preliminary experiment. The level may even be a range of concentrations that is appropriate for the biological system.

Provide enough replicates A single result is not statistically valid. The same treatment must be applied to many individuals and the experiment must be repeated several times to be confident that the results are reliable.

Make predictions about the outcome of your experiment Predictions provide a sense of direction during both the design stage and the data analysis stage of your experiment. For each treatment and control group, predict the outcome of the experiment if your hypothesis is supported. You may also choose to state the null hypothesis, which is that the treatment has no effect on the response.

HYPOTHESIS:	Adding GA to dwarf plants will allow them to grow to the height of normal, wild-type plants.
TREATMENT GROUP:	Dwarf plants + GA
PREDICTION IF HYPOTHESIS IS SUPPORTED:	Dwarf plants will grow as tall as wild-type plants + no GA.
NULL HYPOTHESIS:	Dwarf plants will not grow to the height of wild-type plants.
NEGATIVE CONTROL:	Dwarf plants + no GA
PREDICTION:	Dwarf plants will be short.
POSITIVE CONTROL:	Wild-type plants + no GA
PREDICTION:	Wild-type plants will be tall.

Record data

Scientists record procedures and results in a laboratory notebook. The type of notebook (bound or loose leaf, with or without duplicate pages) may be prescribed by your instructor or the principal investigator of the research lab. More important than the physical notebook, however, is the detail and accuracy of what's recorded inside. For each experiment or study, include the following information:

- Investigator's name
- The date (month, day, and year)
- The purpose
- The procedure (in words or as a flow chart)
- Numerical data, along with units of measurement, recorded in well-organized tables
- Drawings with dimensions and magnification, where appropriate. Structures are drawn in proportion to the whole. Parts are labeled. Observations about the appearance, color, texture, and so on are included.
- Graphs, printouts, and gel images
- Calculations
- A brief summary of the results
- Questions, possible errors, and other notes

When deciding on the level of detail, imagine that, years from now, you or someone else wants to repeat the experiment and confirm the results. The more information you provide, the easier it will be to understand what you did, what problems you encountered, suggestions for improving the procedure, the results you obtained, how you summarized the data, and how you reached your conclusions.

Summarize numerical data

The raw data in lab notebooks are the basis for the results published in the primary and secondary literature. Published results, however, usually represent a *summary* of the raw data by the author, who is both knowledgeable about the subject and intimately familiar with the experiment. We rely on the author's experience and integrity to reduce the original data to a more manageable form that is an honest representation of the phenomenon and which lends itself to interpretation.

How the author presents data in the Results section depends in part on the scope of the question asked at the beginning. Broad questions about a population involve **statistical inference**, whereby results from a sample or subset of the population are applied to the whole. Because a different sample may produce different results, the author includes a statement about the reliability of his or her conclusions using appropriate statistical language. On the other hand, narrower questions about a specific situation may be answered from the data at hand. For example, questions such as "Which fraction of a purification procedure contains the most enzymatic activity?" or "Which medium produces the highest concentration of bacteria?" can be answered from the collected data and require no inference about a larger population. When the data are consistent from one experiment to the next, scientists gain confidence that their conclusions are valid.

When *you* are given the task of summarizing the raw data, first distinguish between trustworthy and erroneous data. Erroneous data include results obtained by dubious means, for example, by not following the procedure, using the equipment improperly, or making simple arithmetic errors. Trustworthy data include results obtained legitimately, but which may still have quite a bit of unexplained variability. If time permits, repeat the experiment to determine possible sources of variability and make changes in the procedure if necessary.

Once you've identified which data are reliable, graph them. It is easier to spot patterns and outliers on a graph than in a table. Furthermore, graphs are used to check assumptions for certain statistical methods. Use bar graphs when one of the variables is categorical (i.e., it has no units of measurement). Use scatterplots and line graphs when both variables are quantitative. Look for an overall trend as well as deviations from the trend. Reduce the data by taking the average (mean) and express variability, where appropriate, in terms of standard deviation or standard error. Never eliminate data without a good reason.

Analyze the data

Once you have a visual summary of the raw data, look for relationships between variables. Do the results match the predictions if the hypothesis is supported? If so, then compare your results to those in the primary references you consulted to develop your hypothesis in the first place. Comparable data from different studies help researchers gain assurance that their conclusions about a particular phenomenon are valid. When analyzing data, however, do not let your predictions affect your objectivity. Do not make your results fit your predictions—instead, modify your hypothesis to fit your results. What is learned from a negated hypothesis can be just as valuable as what is learned from a "successful" experiment.

Keep in mind that there may be no difference between the control and the experimental treatments. If there was no difference, say so, and then try to develop possible explanations for these results.

Try to explain the results

Once you have summarized and analyzed the data, you are ready to develop possible explanations for the results. You previously found information on your topic when you developed your hypothesis. Return to this material to try to explain your results. Do your results agree with those of other researchers? Do you agree with their conclusions? If your results do not agree, try to determine why not. Were different methods, organisms, or conditions employed? What were some possible sources of error?

You should realize that even some of the most elementary questions in biology have taken hundreds of scientists many years to answer. One approach to the problem may seem promising at first, but as data are collected, problems with the method or other complications may become apparent. Although the scientific method is indeed methodical, it also requires imagination and creativity. Successful scientists are not discouraged when their initial hypotheses are discredited. Instead, they are already revising their hypotheses in light of recent discoveries and planning their next experiment. You will not usually get instant gratification from applying the scientific method to a question, but you are sure to be rewarded with unexpected findings, increased patience, and a greater appreciation for the complexity of biological phenomena.

Revise original hypotheses to take new findings into account

If the data support the hypothesis, then you might design additional experiments to strengthen the hypothesis. If the data do not support the hypothesis, then suggest modifications to the hypothesis or use a different procedure. Ideally, scientists will thoroughly investigate a question until they are satisfied that they can explain the phenomenon of interest.

Share findings with other scientists

The final phase of the scientific method is communicating your results to other scientists, either at scientific meetings or through a publication in a journal. When you submit a paper to refereed journals, it is read critically by other scientists in your field, and your methods, results, and conclusions are scrutinized. If any errors are discovered, they are corrected before your results are communicated to the scientific community at large.

Poster sessions are an excellent way to share preliminary findings with your colleagues. The emphasis in poster presentations is on the methods and the results. The informal atmosphere promotes the exchange of ideas among scientists with common interests. See Chapter 7 on how to prepare a poster.

Oral presentations are different from both journal articles and poster sessions, because the speaker's delivery plays a critical role in the success of the communication. See Chapter 8 for tips on preparing and delivering an effective oral presentation.

Go to the **COMPANION WEBSITE** • **sites.sinauer.com/knisely5e**
for samples, template files, and tutorial videos

Chapter 2

Developing a Literature Search Strategy

The development of library research skills is an essential part of your training as a biology student. A vast body of literature is available on just about every topic. Finding exactly what you need is the hard part.

In biology, sources are divided broadly into primary and secondary references. **Primary references** are the research articles, dissertations, technical reports, or conference papers in which a scientist describes his or her original work. Primary references are written for fellow scientists—in other words, for a specialized audience. The objective of a primary reference is to present the essence of a scientist's work in a way that permits readers to duplicate the work for their own purposes and to refute or build on that work.

Secondary references include encyclopedias, textbooks, articles in popular magazines, and information posted on the websites of professional societies, government agencies, and other scientific organizations. Secondary references are based on primary references, but they address a wider, less-specialized audience. In secondary references, there is less emphasis on the methodology and presentation of data. Instead, the results and their implications are described in general terms for the benefit of non-specialist readers.

You will delve into the biological literature when you write laboratory reports, research papers, and other assignments. Although secondary references provide a good starting point for your work, it is important to be able to locate the primary sources on which the secondary sources are based. Only the primary literature provides you with a description of the methodology and the actual experimental results. With this information, you can draw your own conclusions from the author's data.

Although initially it may be difficult to read primary literature, it will become easier with practice, and the rewards are well worth it. One benefit of *reading* research articles is that you will become a better *writer*. Through reading, you become familiar with the writing style and overall

TABLE 2.1 Databases and search engines used to find scholarly information in the biological sciences

Database or Search Engine	Description
AGRICOLA	Produced by the US Department of Agriculture's National Agricultural Library, this database contains citations for journal articles, monographs, government publications, patents, and other types of publications in the field of agriculture and related areas.
Biological Abstracts	Considered the most comprehensive database in the area of biology and the life sciences, it provides abstracts and citations to journal literature.
Biological Science (ProQuest)	Indexes scholarly and trade journals, books, conference proceedings, government publications, and other publication types for a wide range of areas in the life sciences.
BioOne	A journal collection of full-text, peer-reviewed articles in biology and the environmental sciences. Most of the journals are published by small scientific societies, other not-for-profits, and open access publishers.
Google Scholar	A Web search engine for scholarly literature across many disciplines, languages, and countries. Includes not only journal articles, but also material from websites of universities, scientific research groups, and professional societies; conference proceedings; court opinions and patents; and preprint archives (**preprints** are manuscripts circulated because they contain current information, but they have not yet been peer reviewed). Articles in the popular press, book reviews, and editorials are not included.

structure of research articles, so that you have a model when you write your own lab reports. Another benefit is that you learn how scientists approach a problem, design experiments to test hypotheses, and interpret their results to arrive at their conclusions. Emulating their writing style may help you improve your critical thinking skills. A further benefit of reading the primary literature is getting to know the scientists who work in a particular subdiscipline. You may discover that you are sufficiently interested in a subdiscipline to pursue graduate work with one or more of the authors of a journal article.

How do you find primary references that are directly relevant to your topic? The fastest and easiest way is to search article databases. **Article databases** contain a pre-screened collection of scholarly information, not web pages that anyone could have created. Article databases are owned by companies or organizations that employ experts to read scholarly arti-

TABLE 2.1 *(continued)*

Database or Search Engine	Description
JSTOR	Developed as a digital archive of core scholarly journals, this database searches the full text of core journals in a variety of disciplines including biology and ecology. Coverage begins with the first issue of each journal. However there is a gap, typically from 1 to 5 years, between the most recently published issue and when it appears in JSTOR.
NCBI (National Center for Biotechnology Information)	A division of the US National Library of Medicine. Produces searchable databases on nucleotide and protein sequences, protein structures, complete genomes, taxonomy, and other molecular biology information.
PubMed	Produced by the US National Library of Medicine, PubMed is the public access version of MEDLINE, the premier database for medicine and related fields. It contains abstracts and citations to the worldwide journal literature and books.
ScienceDirect	Provides access to journal articles and books published by Elsevier. Although multidisciplinary, most references are in the areas of science, medicine, and engineering.
Scopus	A database of scientific information resources, including journal articles, books, and conference proceedings. Almost 40% of the records are pre-1996; over 60% of the records are post-1995.
Web of Science	An interdisciplinary database for peer-reviewed articles from core journals in many subject areas. The "Times Cited" and "Cited References" searches allow you to identify more recent and older articles, respectively, which cite a particular author or work.

Source: Kathleen McQuiston, Research Services Librarian, Library and Information Technology, Bucknell University (2016 Oct 23) and respective database or search engine websites.

cles and then enter information about the articles into the database. To find scholarly information on a particular topic, instead of "googling" the entire Web, you will typically search one or more databases.

Most databases (PubMed being the notable exception) are by subscription. Companies that own these databases sell licensing agreements to university libraries and other institutions. If you are affiliated with such a university or institution, then you can use fee-based databases for free. On the other hand, search engines such as Google Scholar, which scan the Web for scientific information, are free and available to the general public. Table 2.1 describes some of the databases and scholarly search engines that you may have access to. Many of these databases have apps that can be installed on your mobile devices.

Most of this chapter describes how to find primary references using databases. If you do not have access to these databases, however, you can still locate references the old-fashioned way. This method involves building a bibliography from sources cited in books, journal articles, and other literature. Books such as *Annual Reviews* are considered secondary references, but the Literature Cited section in review articles often is an excellent source of primary references. Building a bibliography without the use of a database is laborious and time-consuming, but the end result is often the same. An advantage of using this old-fashioned method is that you may find older, seminal papers that may not be indexed in databases. If your assignment requires a thorough search of the literature, you will most likely use a combination of database and manual searches. Don't forget about your human resources—seek assistance from your reference librarian during all stages of your research project.

Databases and Search Engines for Scientific Information

Familiarize yourself with the databases and search engines recommended by your professor or a reference librarian and which are available through your academic library. All of the databases have some overlap in terms

TABLE 2.2 Comparison of features of selected biology databases and search engines

	Biological Abstracts	Google Scholar
Resource type	Database	Search engine
Access (free or fee-based)	Fee-based (usually institutional subscription)	Free
Years covered	1926 to present	Unknown
Sources retrieved	Journal articles	No information provided, but retrieves journal articles, books, preprints, abstracts, technical reports, and other electronic media
Content (number of journals indexed)	Searches more than 5,200 journals in the life sciences	Unknown
Reliability (peer-reviewed materials)	Most journals are peer-reviewed	Unclear whether all journal articles are peer-reviewed

Source: Kathleen McQuiston, Research Services Librarian, Library and Information Technology, Bucknell University (2016 Oct 23) and respective database or search engine websites.

of the journals they index, but there are also unique listings. Results may also vary depending on subject and publication year.

Comparison of databases

One of the great things about electronic databases is that they are continually updated and improved, giving you access to the most current scientific information available on the Internet. But with so many choices and so little time, what's the best strategy for tracking down a few good primary journal articles for your topic? The answer to this question depends on who you ask, how comprehensive your research needs to be, the subject matter, and personal search preferences. Nonetheless, knowing a little about the strengths and weaknesses of some of the major databases and search engines may help guide your strategy (Table 2.2).

Librarians and scholars interested in information technology have published a number of recent papers on this topic (see, for example, Harzing and Alakangas [2016], Hodge and Lacasse [2013], and Moed *et al.* [2016]). While these published comparisons are as transient as the databases they describe, it is nonetheless instructive to look at some of the data.

TABLE 2.2 *(continued)*

PubMed	Web of Science	Scopus
Database	Database	Database
Free	Fee-based	Fee-based
Generally 1946 to present	1900 to present	1823 to present
Journal articles, literature reviews, clinical trials	Journal articles and conference proceedings	Journal articles, books, patents, and conference proceedings
Biomedical journal citations and abstracts from over 5,000 journals	Searches more than 12,000 journals (all disciplines) and 3.8 million conference proceedings	Searches more than 21,500 journals from more than 5,000 international publishers from all subject areas
Most journals are peer-reviewed	All journals are peer-reviewed	All journals are peer-reviewed

Google Scholar Google Scholar was introduced by Google in 2004. Its strengths are name recognition, a simple query box, and the fact that it's free. In terms of content, Google Scholar is thought to provide greater access to older records and to material not easily located through conventional channels such as publishers' websites. Its web-crawling robots use an algorithm to determine what is "scholarly" based on information provided by authors and publishers on their websites. Some of Google Scholar's weaknesses include the scope of its coverage (it finds too much information), uncertainty about the scholarly value and currency of some of the records, and the sorting of records according to how relevant they are (based in part on how often they were cited). The search results cannot be sorted by date, but a custom date range can be selected.

PubMed PubMed is *the* most recommended database for researchers in medicine who require advanced search functions. Like Google Scholar, PubMed is free and its advanced search feature makes it possible to limit searches by author, publication, and date. PubMed provides a variety of options to retrieve only certain formats (full text, free full text, or abstract), types of article (clinical trial, review, clinical conference, comparative study, government publication, etc.), language, and content (journal group, research topic, humans or animals, gender, and age). Another feature that makes PubMed so powerful is its search algorithm, which is based on concept recognition, not letters or words. Every document indexed for PubMed has been read by experts, who tag the document with controlled vocabulary (Medical Subject Headings or MeSH) that accurately describes the paper's content. "False hits" due to homographs (e.g., swimming pool rather than gene pool) are thus eliminated in PubMed searches. Furthermore, MeSH solves the problem of ambiguity concerning scientific and popular names of organisms, synonyms, and variations in British and American spelling.

Web of Science Web of Science is fee-based, so you may only have access to this database if your university has a subscription. Web of Science covers a larger period of time than either PubMed or Google Scholar. It has depth and scope and is useful for finding information on topics of an interdisciplinary nature. The greatest benefit of this database, however, lies in the fact that once you have found a good journal article, you can expand your bibliography quickly based on common references. With Web of Science, it is possible to search *forward* in time to find more recent papers that have cited the paper of interest. It is also possible to search *backward* to find papers cited by authors of the paper of interest.

Scopus Like Web of Science, Scopus has a tremendous scope in terms of years covered and sources retrieved, and it is fee-based. Scopus, like all of the databases in Table 2.2, has an advanced search feature, provides links to full-text articles, and allows references to be exported to reference management software (see p. 24). In addition, graduate students and career researchers will find the email alerts feature of these databases handy for staying current with the literature. When registering for email alerts, you can enter keywords that are relevant to your research. When a new article containing these keywords appears, the database administrator will send you an email alert.

Database Search Strategies

Finding just the right journal articles on your topic can be a daunting task. This section will help you get started.

Understand your topic

A productive and efficient search begins with a **basic understanding of your topic**. If you don't even know where to start, look up the most specific term you can come up with in the index of your textbook. Open the book to the pages that contain this term. Read the chapter subheadings and the chapter title to learn how this term fits into the bigger picture. Read the relevant pages to find out what subtopics are associated with this term.

Your library's stacks are another good place to find general information. Search the library's catalog to locate a book on your topic. Write down the call number and find this book on the shelf. Browse the titles of other books in the vicinity. Because the Library of Congress cataloging system groups books according to topic, you can often find additional sources shelved nearby.

Encyclopedias and dictionaries may also help you clarify your topic. Check your library's homepage for references that you may have access to, both electronic and printed sources. Websites such as Wikipedia (http://www.wikipedia.org), WebMD (http://www.webmd.com), and others may be a good place to start, but evaluate Internet sources critically. Whereas journal articles and books have undergone a rigorous review process, information on the Web may not have been checked by any authority other than the owner of the website.

A first step in evaluating a website's reliability is to look at the ending of the URL address (Table 2.3). Is the sponsor of the website a company or organization that is more interested in trying to sell a product or idea than in presenting factual information? To become a savvy website evaluator, check out the tips on your library's homepage or take one of the tutorials listed in the Bibliography.

TABLE 2.3 Identifying sponsors of sites on the World Wide Web

Type of Web Page	Purpose	Ending of URL Address	Examples
Informational	To present (factual) information	.edu, .gov	Dictionaries, directories, information about a topic
Business/ marketing	To sell a product	.com	Carolina Biological Supply, Leica
Advocacy	To influence public opinion; to promote the exchange of knowledge and provide resources for its members	.org	Sierra Club, Association for Biology Laboratory Education
News	To present very current information	.com	CNN, *USA Today*

Source: Alexander and Tate (c1996–2005).

Define your research goals

Once you have a basic understanding of your topic, try to **define your research goals** with statements such as

- I would like to compare or contrast methods.
- I'm looking for a cause-and-effect relationship.
- I want to understand more about a process.
- I am interested in how an organism carries out a particular function (e.g., obtains nutrients, reproduces, moves, responds to changes in its environment).

Subdivide your topic into concepts

Once you have formulated the goals for your topic, start **defining smaller concepts**. For example, if the methods you wish to compare have to do with measuring the amount of protein in a sample, then one of the concepts is protein quantification. Another concept would include the specific names of protein quantification methods, such as Lowry, biuret, Bradford, BCA, and so on. A third concept might relate to the types of protein samples that were analyzed.

Another way to organize concepts related to your topic is to use PubMed's Medical Subject Headings (MeSH) database, a kind of thesaurus for the life sciences. Words entered in the search box are translated

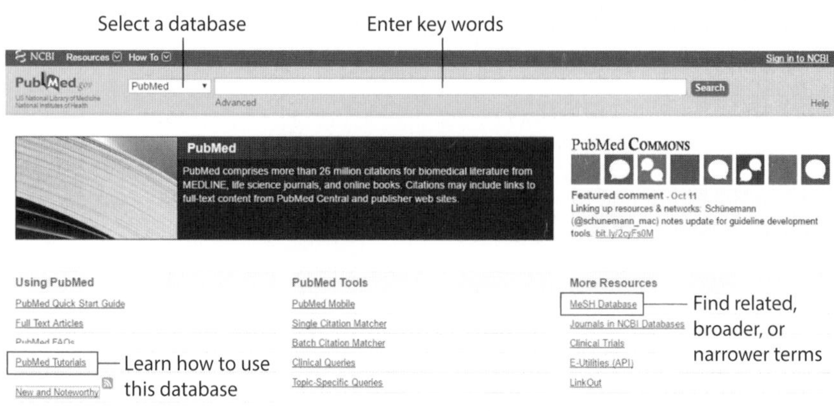

Figure 2.1 PubMed home page provides tutorials and options for searching different databases.

into standardized descriptors, which are then listed in a hierarchy of headings and subheadings.

Let's say, for example, that you would like to find concepts related to the topic "How do *Tetrahymena* move?" Go to the PubMed home page (https://www.ncbi.nlm.nih.gov/sites/pubmed) and select **MeSH Database** under **More Resources** (Figure 2.1). A search for the term *motility* lists *cell movements* as the first result (Figure 2.2); clicking this descriptor opens a page that gives a definition of *cell movement* (not shown in Figure 2.2), entry terms, and the MeSH tree for this concept. The headings below *cell movements* in the tree are narrower concepts and the headings above are broader. Write down the entry terms and headings that are relevant to your topic. While the entry terms are automatically searched in databases that use MeSH, they may be useful keyword alternatives in databases or search engines that do not.

Choose effective keywords

Effective keywords are neither too broad nor too narrow in scope. Keywords that are too broad will retrieve an unmanageable number of articles that, for the most part, are not relevant to your topic. On the other hand, keywords that are too specific may not get any results. For each concept in your topic, therefore, try to come up with moderately specific terms, synonyms, and related descriptors (Figure 2.3). Consider different word endings (photosynthetic versus photosynthesis), abbreviations (*HIV* for *human immunodeficiency virus*), and alternative spellings (American versus British English). **Avoid vague terms** like *effect* and *relationship between*.

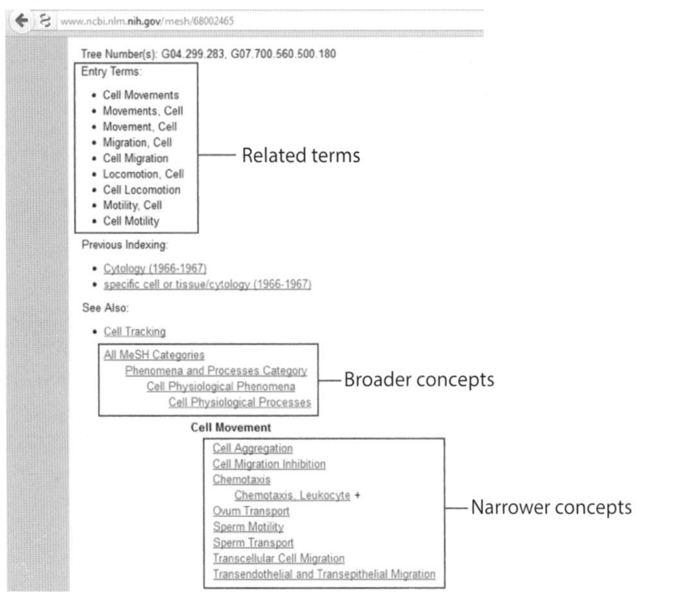

Figure 2.2 MeSH tree for the concept *cell movement*. The related Entry Terms shown above the tree are automatically included in a search for the phrase *cell movement* in databases such as PubMed, which use MeSH.

Connect keywords with the operators and, or, *or* not

After you have generated a list of keywords, select two or more and combine them in a search string using operators such as *and*, *or*, or *not*.

- When the word *and* is used between keywords, the references must have both words present. This connector is a good way to limit your search.
- When the word *or* is used, the references must have at least one of the search terms. This connector is a good way to expand your

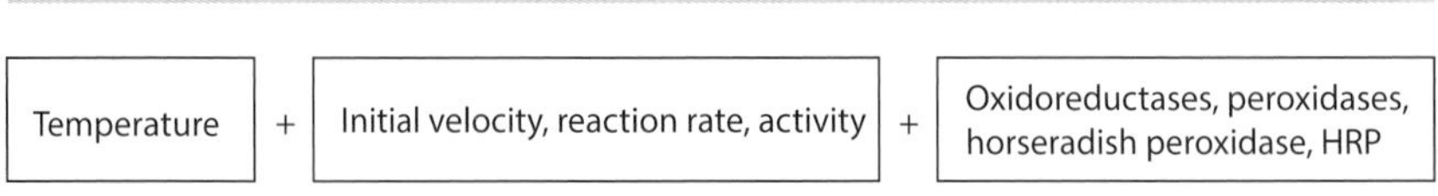

Figure 2.3 Possible keywords generated from concepts related to the topic "How temperature affects the initial velocity of peroxidase, an enzyme isolated from horseradish."

search. For example, the search string *biuret or Bradford* turned up over 9,984 hits in PubMed, while *biuret and Bradford* resulted in only 19 (another search may not result in the same numbers, but the difference would likely be of the same magnitude).

- When the word *not* is used, then the references should not contain that particular keyword. This connector is another way to limit your search.

Use truncation symbols for multiple word endings

Truncation is a method for expanding your search when keywords have multiple endings. For example, many words related to the concept of temperature begin with *therm,* such as thermoregulation, thermoregulatory, thermy, and thermal. Rather than writing a lengthy search string containing all of these terms, simply type *therm* followed by a wildcard symbol like *, ?, or #. The appropriate truncation symbol can be found in the Help menu of the database you are searching. Google Scholar uses stemming technology instead of truncation, whereby it automatically searches for variations in word endings for the given keyword.

Search exact phrase

When the keyword is a phrase, the search engine typically searches for adjacent words in order. Unfortunately, the search results may also include "false hits" in which the words are separated. When it's important to search an exact phrase, use quotation marks. For example, type *"RNA polymerase"* instead of *RNA polymerase.*

Use the same keywords in a different database or search engine

If you are not having any success with different keyword combinations in one search engine or database, try a different one. Google Scholar searches the entire Web and may find links to published journal articles on scientists' homepages or course websites. These media are not included in PubMed or Web of Science database searches. Take advantage of the resources at your disposal. Remember that once you find that one good journal article, it will be much easier to find others (see the section "Finding related articles").

Evaluating Search Results

After you type a keyword string into the search box, the search engine goes to work. The result is a page that lists the records by publication date

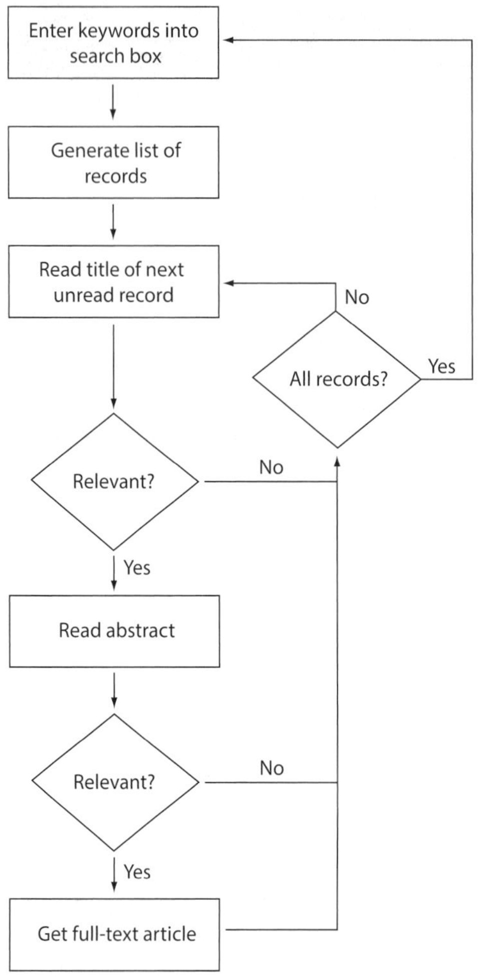

Figure 2.4 Evaluating database or search engine results is an iterative process.

(most recent first), relevance, or another criterion. Each journal article record contains the article title, the authors' names, the name of the journal, the volume and issue numbers, the pages, and the publication date. Based on the title, decide if you want to read the abstract. After having read the abstract, decide whether you want to read the entire paper. This iterative process is summarized in Figure 2.4.

The results pages for Web of Science and PubMed are formatted slightly differently, but both contain the same basic information about the journal articles (Figure 2.5). You will need this information when you

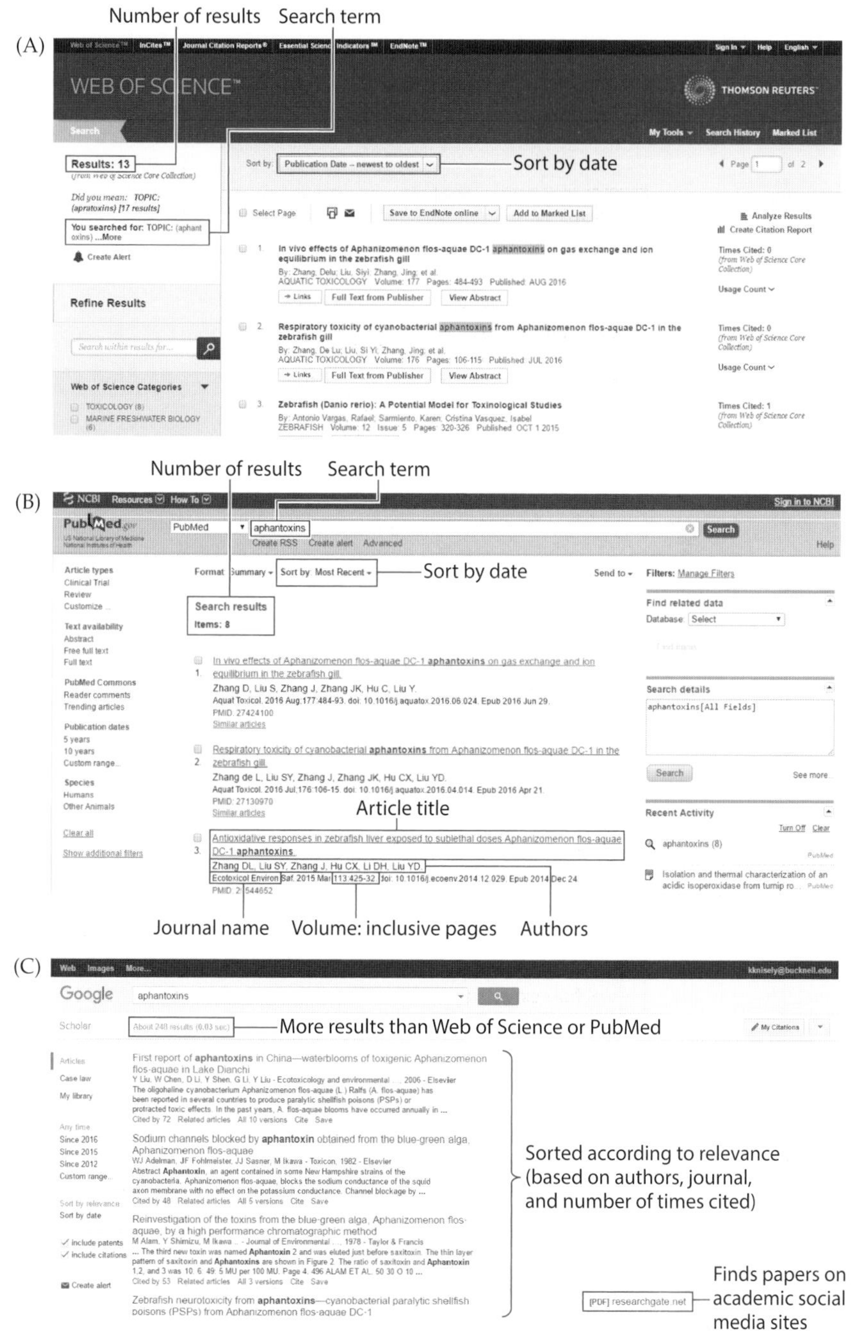

Figure 2.5 The results page from (A) Web of Science, (B) PubMed, and (C) Google Scholar for the keyword phrase *aphantoxins.*

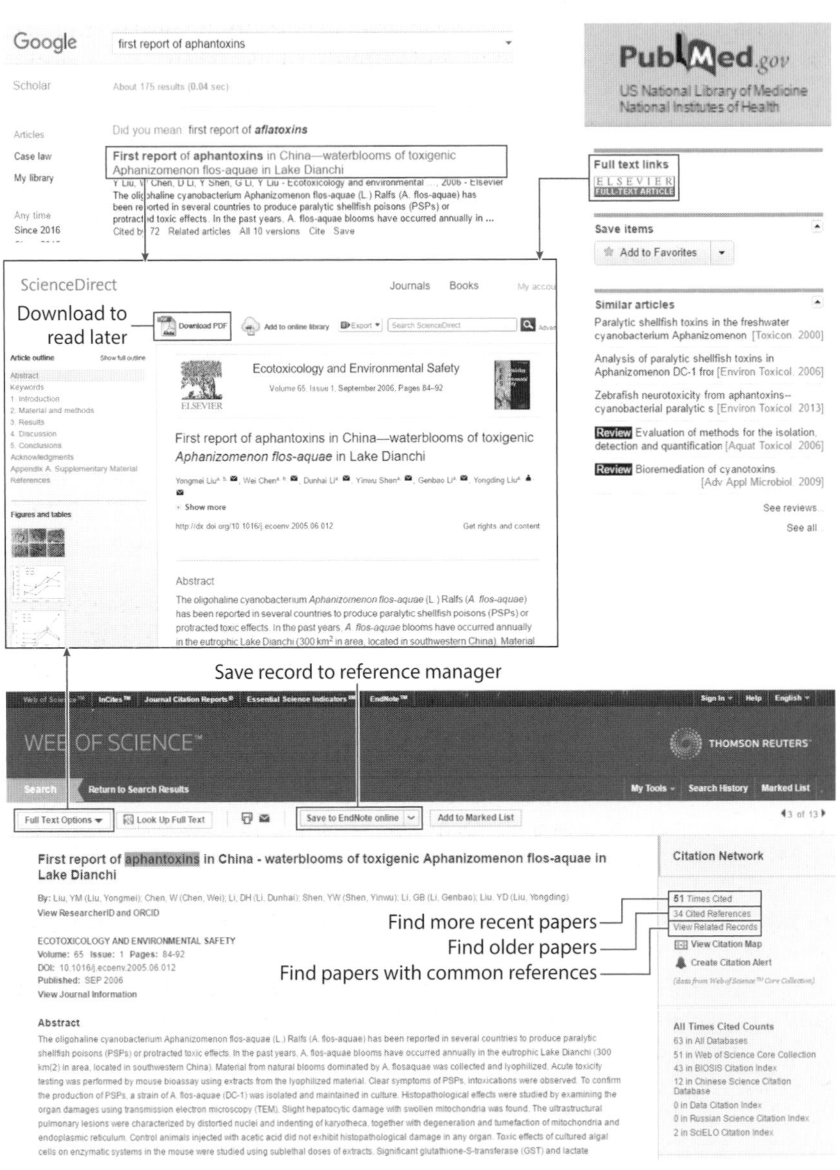

Figure 2.6 Detailed record from Web of Science showing the abstract, a link to the full-text article, and links to related articles. The full-text article is stored in ScienceDirect, a repository of academic journals and ebooks managed by Elsevier. The corresponding links from Google Scholar and PubMed lead to the same full-text article. The citation information can also be saved to a reference manager such as EndNote, Mendeley, RefWorks, or Zotero.

cite the article in your lab report or research paper (see the section "Documenting Sources" in Chapter 4). Google Scholar, on the other hand, lists the authors, journal name, year, and publisher followed by an excerpt of the abstract or passage where the keywords are used.

Skim the titles of the first 20 records. If the titles seem to be unrelated to your topic, start a new search with different keywords using the strategies described previously (see the section "Choose effective keywords"). If a title seems promising, click it to open a page that contains the abstract (Figure 2.6). Based on the title and the abstract, decide whether or not you want to read the entire article. In Google Scholar, clicking the title takes you directly to the source text.

Finding related articles

Once you have found a good article, Web of Science makes it easy to find related articles. In the **Times Cited** section, there is a list of more recent papers that cite this article (see Figure 2.6). Clicking on one of these titles opens a new page that displays the abstract of the more recent paper. In the **Cited References** section, you can view the references listed in the article. Browsing the list allows you to find related papers with a slightly different focus. In the **View Related Records** section, papers are listed, which cite references that were also cited in the article. Common references indicate that the authors were pursuing a similar research topic. PubMed also offers a **Similar Articles** option (see Figure 2.6).

Finding review articles is the equivalent of hitting the mother lode. **Review articles** are secondary references that summarize the findings of all major journal articles on a specific topic since the last review. You can find background information, the state of current knowledge, and a list of the primary journal articles authored by scientists who are working on this topic. If you are unable to find a relevant review article in a database, go directly to the Annual Reviews website (www.annualreviews.org) and search for your topic. If you find a promising review article on this website, you may be able to obtain a copy through your academic library.

Most of the article databases and search engines also have an advanced search option. Advanced search makes it possible for you to limit your search by specifying one or more authors, publication years, journals, and other criteria.

Obtaining full-text articles

If the title and the abstract of an article sound promising, you will want to obtain the full-text article. Web of Science, PubMed, and Google Scholar all have links to full-text articles that you can download as a PDF (see Figure 2.6). Some publishers also offer an HTML option. PDFs preserve formatting, while HTML files contain hyperlinks that make it easy to access other references. Save the full-text article to your computer or cloud storage space to read later. Copy the URL and write down the download date, because you may need this information when citing the source.

While the abstract is usually free, some publishers charge a fee to access the full-text article. Fortunately, academic libraries and institutions often purchase subscriptions so that faculty, staff, and students can obtain many electronic journal articles for free. If your library does not have a subscription and you are not in a hurry to get the article, you may be able to use interlibrary loan. **Interlibrary loan** is a way for a library to borrow or obtain materials that it does not own from another library or organization.

Managing References (Citations)

Reference management software makes it possible to

- Build your own collection of references from database searches.
- Insert citations into a paper.
- Format both the in-text reference and the end reference according to the style specified by your instructor. You can select from hundreds of styles, including the familiar CSE, APA, MLA, and Chicago styles. If you are submitting your paper to a journal, RefWorks even offers styles for specific journals.

Some of these products are free to everyone (e.g., Mendeley and Zotero) and others are free as long as you are affiliated with a subscribing institution (e.g., RefWorks and EndNote).

Many scientists and other scholars rely heavily on reference management software to organize all of their references. Students will appreciate the convenience and ease of use of these programs as well. The following instructions for RefWorks are intended simply to make you aware of the possibilities. If you like what you see, ask your librarian if you can access something similar at your school.

ProQuest RefWorks

Create an account Go to the RefWorks login page found at https://refworks.proquest.com. Enter your institution's credentials and create an account. You will receive an email confirmation to complete the process. To learn the basics of RefWorks in 20 minutes, watch the helpful YouTube tutorials available at https://www.youtube.com/channel/UCzmTj_AGeY59VoNv-0SvcCg.

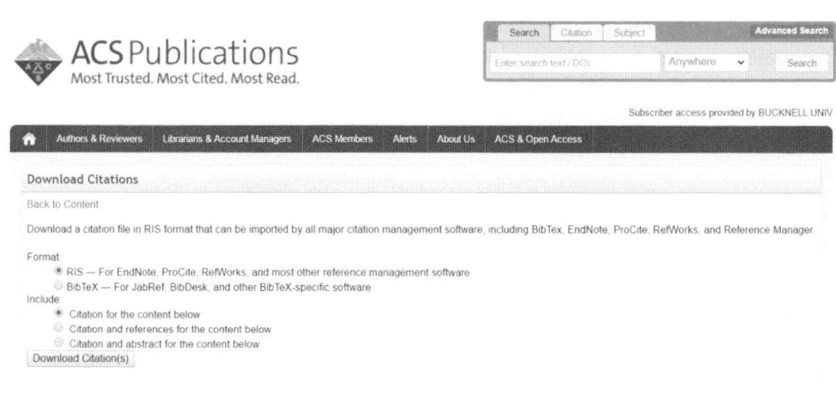

Figure 2.7 The Download Citations dialog box is used to specify the format (use RIS for RefWorks) and whether the abstract or any references are to be included with the citation.

Download citation into RefWorks

1. After you have found a reference that seems useful, click the **Cite** or **Export** or **Download to citation manager** link in the database or on the publisher's website.
2. In the Download Citations dialog box, click the desired content: **Citation only**, **Citation and references**, or **Citation and abstract** (Figure 2.7). Click **Download Citation(s)**. The citation file will be downloaded to your computer in RIS format.
3. In RefWorks, click **+ (Add a Reference) | Import References** (Figure 2.8). Under **Import from a file**, click the "select a file from your computer" hyperlink, navigate to the .ris file, and click **Open**. It doesn't seem to matter that Abbott Labs is the default, even though the file was downloaded from a different database.
4. The dialog box will notify you that 1 reference was imported. Click **Last Imported** to see the details of the reference.
5. References are easier to find when they are assigned to folders. To create a new folder, click **My Folders | Add a folder** (see Figure 2.8). Then drag the imported reference into the new folder or into an existing one.

Download Write-N-Cite or RefWorks Citation Manager Before you can insert the references you saved in RefWorks into a paper, you have to download an add-in to your computer.

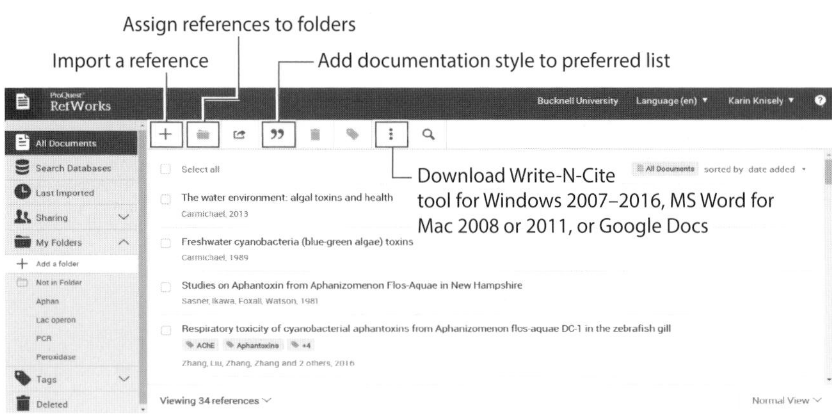

Figure 2.8 In ProQuest RefWorks, downloaded references can be added to folders to facilitate reference management. There are hundreds of documentation styles to choose from, including CSE 8th edition Name-Year and Citation-Sequence systems. The Write-N-Cite tool is downloaded from the **⋮ (More) | Tools** menu. This tool makes it easy to insert in-text references in Word documents and then build a references list (or bibliography) at the end.

- If you have Microsoft Word 2007 through 2016 for Windows, download either Write-N-Cite or RefWorks Citation Manager. To download Write-N-Cite, click **⋮ (More) | Tools** (see Figure 2.8), scroll down to Cite in Microsoft Word, and click the **Download & install** button. Follow the prompts to complete the installation. Make sure no Word documents are open during the installation. To download RefWorks Citation Manager in Word 2016, see the instructions for "Word for Mac 2016" below.
- If you have Word for Mac 2008 or 2011, navigate to Cite in Microsoft Word as above, except click **Other Windows and Mac Versions** to get Write-N-Cite.
- Word for Mac 2016 users only have one option, namely, to download RefWorks Citation Manager. In Word, click **Insert | Store**, and type "refworks" in the search box. Click **Add** to begin the installation.
- The Google Docs add-in can be installed from the **⋮ (More) | Tools** menu in RefWorks. Google Docs allow you to create and edit MS Office documents online, share your documents, and access them from mobile devices. Their main disadvantages include having fewer features than the MS Office programs downloaded on your computer. In addition, formatting may

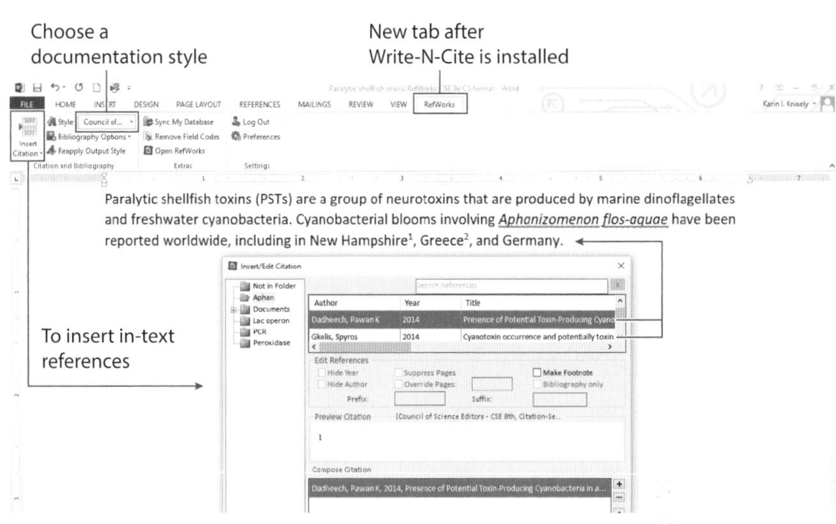

Figure 2.9 To insert a citation in C-S style, position the cursor in the Word document after the period. Click **Insert Citation** and navigate to the folder that contains the reference. After clicking the desired reference and then **OK**, Write-N-Cite inserts a superscripted number.

change when documents created in desktop versions of Office are uploaded to Google Docs.

- Mobile devices. Currently RefWorks supports iPad; support for Office 365 and Android is forthcoming.
- After Write-N-Cite or RefWorks Citation Manager has been installed, open a Word document. You will notice that a new tab, **RefWorks**, has been added to the Ribbon, as shown in Figure 2.9. Click **Settings | Log In**, and enter your login information (for personal computers, this is a one-time procedure). Write-N-Cite syncs your references in RefWorks so that you will be able to insert them in your Word document even when you are not connected to the Internet.

Create an in-text reference and end reference list When you write lab reports and research articles, you will cite the work of others and then, at the end of your paper, list the full references. Many biologists use the Council of Science Editors (CSE) style, which is quite different from the MLA or Chicago Style you may be accustomed to using in the humanities. The CSE recommends the following three systems:

- Citation-Sequence
- Name-Year
- Citation-Name

The Citation-Name system is a hybrid of the other two and will be discussed briefly in Chapters 3 and 4.

CITATION-SEQUENCE (C-S)

In the Citation-Sequence system, in-text references are numbered sequentially and the corresponding full reference is given in a numbered list at the end of the paper.

1. Begin typing your paper in Word. Save the document after you come to a sentence in which you want to cite a reference.
2. Click the down arrow next to **RefWorks | Citation & Bibliography | Style**. The Council of Science Editors styles are not among the top six styles listed. To add CSE styles to the list, open RefWorks in your browser and click the **" (Create Bibliography)** button and then **Create Bibliography**. In the second field from the left on the menu bar, click the down arrow and type "cse" into the search box. Select the **Council of Science Editors – CSE 8th, Citation-Sequence** style.
3. Back in your Word document, click **RefWorks | Extras | Sync My Database** to download the new style to your computer. The six most recently used styles are displayed.
4. Position the cursor *one space after the word* or *after the period* where you want to cite a reference. Click **RefWorks | Citation & Bibliography | Insert Citation**. In the **Insert/Edit Citation** dialog box, navigate to the relevant folder and click the reference that is to be cited (see Figure 2.9). A superscripted number will appear in the Word document.
5. Repeat this process for each reference to be cited.
6. Save the document just before you are ready to generate the end reference list (bibliography). This step is important, because Write-N-Cite will not properly format the in-text reference and the end references list if the document has not been saved.
7. Position the cursor at the end of the document. Click **RefWorks | Citation & Bibliography | Bibliography Options | Insert Bibliography** (Figure 2.10).
8. In your Word document, in-text references are listed sequentially and the information in the end references is in the correct order. Minor editing may be required, but think of the time you'll save by not having to type reference lists!

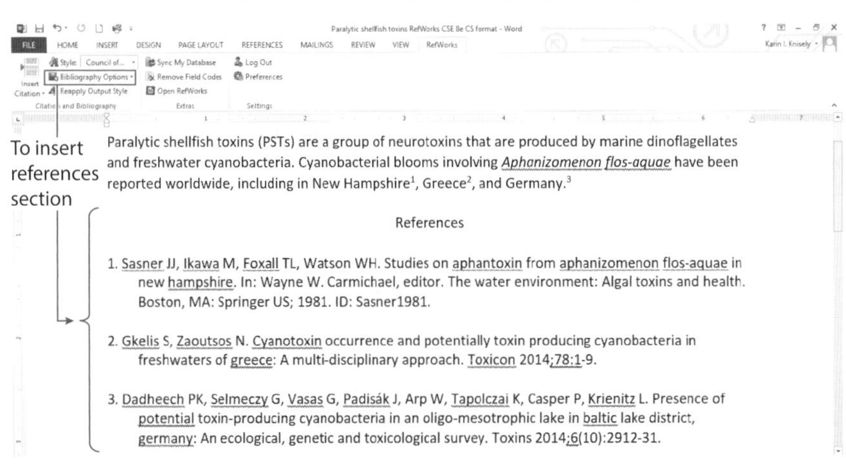

Paralytic shellfish toxins RefWorks CSE 8e CS format - Word
FILE HOME INSERT DESIGN PAGE LAYOUT REFERENCES MAILINGS REVIEW VIEW RefWorks
Style: Council of...
Bibliography Options
Reapply Output Style
Sync My Database
Remove Field Codes
Open RefWorks
Log Out
Preferences
Insert Citation
Citation and Bibliography
Extras
Settings
To insert references section
Paralytic shellfish toxins (PSTs) are a group of neurotoxins that are produced by marine dinoflagellates and freshwater cyanobacteria. Cyanobacterial blooms involving Aphanizomenon flos-aquae have been reported worldwide, including in New Hampshire1, Greece2, and Germany.3
References
1. Sasner JJ, Ikawa M, Foxall TL, Watson WH. Studies on aphantoxin from aphanizomenon flos-aquae in new hampshire. In: Wayne W. Carmichael, editor. The water environment: Algal toxins and health. Boston, MA: Springer US; 1981. ID: Sasner1981.
2. Gkelis S, Zaoutsos N. Cyanotoxin occurrence and potentially toxin producing cyanobacteria in freshwaters of greece: A multi-disciplinary approach. Toxicon 2014;78:1-9.
3. Dadheech PK, Selmeczy G, Vasas G, Padisák J, Arp W, Tapolczai K, Casper P, Krienitz L. Presence of potential toxin-producing cyanobacteria in an oligo-mesotrophic lake in baltic lake district, germany: An ecological, genetic and toxicological survey. Toxins 2014;6(10):2912-31.

Figure 2.10 Final appearance of a sample lab report formatted using the Citation-Sequence system. After clicking **Bibliography Options | Insert Bibliography**, Write-N-Cite generates the end reference list based on the style selected.

NAME-YEAR (N-Y)

In the Name-Year system, the in-text reference is given in the form of author and year. The number of authors determines the format of the citation:

- 1 author: Author's last name followed by year of publication
- 2 authors: First author's last name and second author's last name followed by year of publication
- 3 or more authors: First author's last name followed by the words and others (or *et al.*) and year of publication

The corresponding full references are listed alphabetically at the end of the paper.

1. Begin typing your paper in Word. Save the document after you come to a sentence in which you want to cite a reference.
2. Click the down arrow next to **RefWorks | Citation & Bibliography | Style**. The Council of Science Editors styles are not among the top six styles listed. To add CSE styles to the list, open RefWorks in your browser and click the " **(Create Bibliography)** button and then **Create Bibliography**. In the second field from the left on the menu bar, click the down arrow and type "cse" into the search box. Select the **Council of Science Editors – CSE 8th, Name-Year Sequence** style.

3. Back in your Word document, click **RefWorks | Extras | Sync My Database** to download the new style to your computer. The six most recently used styles are displayed.
4. Position the cursor *one space after the word* or *ahead of the period* where you want to cite a reference. Click **RefWorks | Citation & Bibliography | Insert Citation**. In the **Insert/Edit Citation** dialog box, navigate to the relevant folder and click the reference that is to be cited (see Figure 2.9).
5. Repeat this process for each reference to be cited.
6. Save the document just before you are ready to generate the end reference list. This step is important, because Write-N-Cite will not properly format the in-text reference and the end references list if the document has not been saved.
7. Position the cursor at the end of the document. Click **RefWorks | Citation & Bibliography | Bibliography Options | Insert Bibliography**.
8. In your Word document, in-text references are listed in alphabetical order, and the information in the end references is in the correct order (Figure 2.11). Minor editing may be required, but think of the time you'll save by not having to type reference lists!

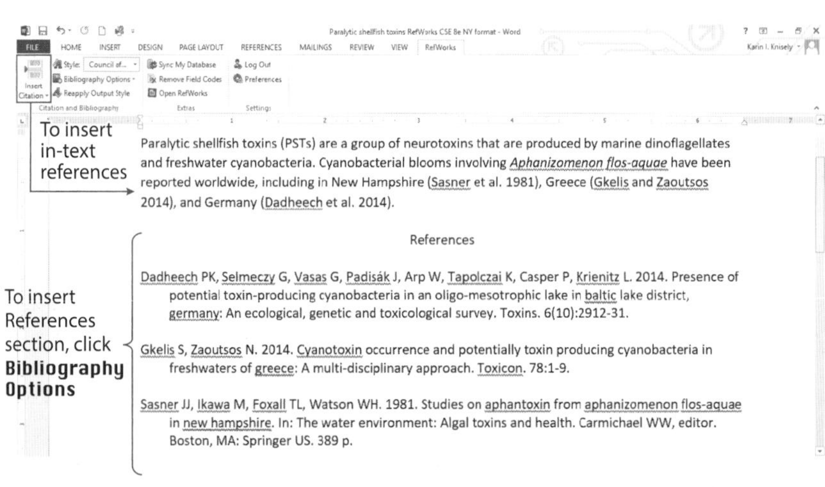

Figure 2.11 Final appearance of a sample lab report formatted using the Name-Year system. After clicking **Bibliography Options | Insert Bibliography**, Write-N-Cite generates the end reference list based on the style selected.

Go to the **COMPANION WEBSITE** • **sites.sinauer.com/knisely5e**
for samples, template files, and tutorial videos

Chapter 3

Reading and Writing Scientific Papers

Reading and writing are two sides of the same coin. We read to acquire knowledge and write to disseminate it. Acquiring knowledge in biology is not just about memorizing facts and practicing lab techniques. Knowledge acquisition is a lifelong process that involves mastering the basics, applying basic information to new situations and problems, and reviewing and reciting the information at regular intervals until it becomes second nature. Disseminating knowledge through written and oral communications requires a certain mastery of the subject matter. Although writing may help you identify gaps in your knowledge, lab reports and other forms of scientific communication are best written *after* you have struggled to understand the material. Your understanding must then be translated into words that convey knowledge clearly and concisely to your readers. Writing well is hard work, but in the long run, good communication skills will open doors to an interesting, challenging, and financially rewarding career in biology.

Types of Scientific Communications

Scientific writing takes many forms. As an undergraduate biology major, you will be asked to write laboratory reports, answer essay questions on exams, paraphrase information from journal articles, and do literature surveys on topics of interest. Third- and fourth-year college students may write research proposals and honors theses and present their work at poster sessions and other venues. Graduate students typically write master's theses and doctoral dissertations and present talks about their research at national and international conferences. Professors write lectures, letters of recommendation for students, grant proposals, reviews of articles submitted for publication to scientific journals by their colleagues, and evaluations of grant proposals. In business and industry, scientific writing may take the form of progress reports, product descriptions, operating manuals, and sales and marketing material. Medical writers research and prepare various kinds of documents and educational materials for healthcare professionals, pharmaceutical companies, and regulatory agencies. Journalists write about science for a broad, non-specialist audience.

Hallmarks of Scientific Writing

What distinguishes scientific writing from other kinds of writing? One difference is the motive. Scientific writing aims to inform rather than to entertain the reader. The reader is typically a fellow scientist who intends to use this information to

- Stay current in his or her field.
- Build on what is already known.
- Improve a method or adapt a method to a different research question.
- Make a process easier or more efficient.
- Improve a product.

A second difference is the style. Brevity, a standard format, and proper use of grammar and punctuation are the hallmarks of well-written scientific papers. The authors have something important to communicate, and they want to make sure that others understand the significance of their work. Flowery language and "stream of consciousness" prose are not appropriate in scientific writing because they can obscure the writer's intended meaning.

A third difference between scientific and other types of writing is the tone. Scientific writing is factual and objective. The writer presents information without emotion and without editorializing.

Scientific Paper Format

Scientific papers, or research papers, are descriptions of how the scientific method was used to study a problem. They follow a standard format that allows the reader, first, to determine initial interest in the paper, second, to read a summary of the paper to learn more, and, finally, to read specific sections of the paper itself for particular details. This format is very convenient, because it allows busy people to scan volumes of information in a relatively short time, and then spend more time reading only those papers that truly provide the information they need.

Almost all scientific papers are organized as follows:

- Title
- List of authors
- Abstract

- Introduction
- Materials and Methods
- Results
- Discussion
- Acknowledgments
- References

This standard structure is sometimes called the IMRD format. **IMRD** is an abbreviation of the core sections of a scientific paper.

The **Title** is a **short, informative description of the essence of the paper**. It should contain the fewest number of words that accurately convey the content. Readers use the title to determine their initial interest in the paper.

Only the names of **people who played an active role** in designing the experiment, carrying it out, and analyzing the data appear in the **List of Authors**.

The **Abstract** is a **summary of the entire paper** in 250 words or less. It contains (1) an introduction (scope and purpose), (2) a short description of the methods, (3) results, and (4) conclusions. There are no literature citations or references to figures in the abstract. If the title sounds promising, readers will use the abstract to determine if they are interested in reading the entire paper.

The **Introduction** concisely states what motivated the study, how it fits into the existing body of knowledge, and the objectives of the work. The introduction consists of two primary parts:

1. **Background or historical perspective on the topic**. Primary journal articles and review articles, rather than textbooks and newspaper articles, are cited to provide the reader with direct access to the original work. Inconsistencies, unanswered questions, or new questions that resulted from previous work set the stage for the present study.
2. **Statement of objectives of the work**. What were the goals of the present study?

The **Materials and Methods** section describes, in full sentences and well-developed paragraphs, **how the experiment was done**. The author provides sufficient detail to allow another scientist to repeat the experiment. Volume, mass, concentration, growth conditions, temperature, pH, type of microscopy, statistical analyses, and sampling techniques are critical pieces of information that must be included. When and where the work was carried out is important if the study was done in the field (in nature), but is not included if the study was done in a laboratory. Conventional labware and laboratory techniques that are common knowledge (familiar to the audience) are not explained. In some instances, it is appropriate to use references to describe methods.

The **Results** section is **where the findings of the experiment are summarized**, without giving any explanations as to their significance (the "whys" are reserved for the Discussion section). A good Results section has two components:

- A *text*, which forms the body of this section
- Some form of *visual* that helps the reader comprehend the data and get the message faster than from reading a lengthy description

In the **Discussion** section, the **results are interpreted** and possible explanations are given. The author may:

- Summarize the results in a way that supports the conclusions.
- Describe how the results relate to existing knowledge (literature sources).
- Describe inconsistencies in the data; this is preferable to concealing an anomalous result.
- Discuss possible sources of error.
- Describe future extensions of the current work.

In the **Acknowledgments** section of published research articles, the authors recognize technicians, colleagues, and others who have contributed to the research or production of the paper. In addition, the authors acknowledge the organization(s) that provided funding for the work as well as individuals who provided non-commercially available products or organisms.

References list the **outside sources** the authors consulted in preparing the paper. No one has time to return to a state of zero knowledge and rediscover known mechanisms and relationships. That is why scientists rely so heavily on information published by their colleagues. References are typically cited in the Introduction and Discussion sections of a scientific paper, and the procedures given in the Materials and Methods section are often modifications of those in previous work.

Styles for Documenting References

The Council of Science Editors (CSE Manual 2014) recommends the following three systems for documenting references:

> **Citation-Sequence System**. In the *text*, the source of the cited information is provided in an abbreviated form as a superscripted endnote or a number in square brackets or parentheses. On the *references pages* that follow the Discussion section, the sources are listed in **numerical order** and include the full reference.

Name-Year System. In the *text*, the source is given in the form of author(s) and year. On the *references pages* that follow the Discussion section, the references are listed in **alphabetical order** according to the first author's last name.

Citation-Name System. This system is a hybrid of the Citation-Sequence and Name-Year systems. In the *text*, the source of the cited information is provided in an abbreviated form as a superscripted endnote or a number in square brackets or parentheses. On the *references pages* that follow the Discussion section, the references are listed in **alphabetical order** according to the first author's last name. The references are then numbered sequentially.

The Name-Year system has the advantage that people working in the field will know the literature and, on seeing the authors' names, will understand the reference without having to check the reference list. This system is more commonly used and generally is preferred. With the Citation-Sequence and Citation-Name systems, for each reference the reader must turn to the reference list at the end of the paper to gain the same information. The Name-Year and Citation-Sequence systems are described in detail on pp. 90–94.

Strategies for Reading Journal Articles

The *way* you read a journal article depends on *why* you are reading it. The strategy described below is a modification of the SQ3R method recommended by learning support staff to improve reading comprehension (see, for example, resources posted by The Teaching and Learning Center at Bucknell University 2016; Counselling Services at the University of Victoria 2016; and links within these websites to other websites). The method itself takes longer to complete than simply sitting down and passively reading the article. However, the investment in time is rewarded by deeper understanding and the ability to recall the information for a longer time afterwards.

Prereading. Also called *previewing* or *surveying*. This method involves skimming a text to get an overview or to find specific information. Prereading is designed for speed, not comprehension. Thus you would preread a journal article to decide if you are sufficiently interested in the content to read the article carefully later. Start by skimming the title, the abstract, the key words (if present), and the first few sentences of the introduction. If the paper seems promising, look for specific information by section. The standard IMRD structure facilitates this process.

Reading for comprehension. Let's assume that you *want* to read this journal article. First preread it to get an idea of the topic, as described above. Then take some time to reflect on what you already know about this topic. If you find that this article is way over your head, you will need to acquire background information. With appropriate background information, you can then engage actively with the content by formulating questions and seeking answers. Write the answers in your own words and then practice saying them out loud from memory. Discuss your newly acquired knowledge with your study group or instructor to improve your understanding. Then share your knowledge. Each of these steps is described below.

Acquire background information on the topic

Papers in scientific journals are written by experts in the field. Because you are not yet an expert, you will probably find it difficult to read and understand journal articles. Even experts may read journal articles several times before they understand the methodology and the implications of the findings.

For convenience, you may start looking for background information on a topic by entering key words in Google, Wikipedia, or even YouTube. However, these websites should not be considered authoritative sources for academic work. A better choice may be your textbook, written by scientists and reviewed by other scientists before publication. Because textbook authors generally write for a student audience, not a group of experts, your textbook is likely to be easier to read than the primary literature. See "Strategies for Reading your Textbook" on pp. 39–42 for ways to read biology textbooks efficiently.

Formulate questions

Active reading means reading with a purpose. Scientists read journal articles specifically to acquire the most up-to-date knowledge about a topic directly from the researchers who did the work. In other words, like those scientists, you are reading a journal article to find answers to specific questions. So before you begin reading, make a list of questions. The following questions, divided according to section, will help you read journal articles with focus.

For the introduction The structure of the introduction is broad to specific. The first few sentences are aimed at attracting reader interest, and the topic is introduced in general terms. Subsequent sentences narrow down the topic, setting the stage for the specific goals of the study, which are usually stated in the last few sentences.

- What is the general topic of this paper?
- What aspect of this topic is being studied?
- What was already known about the specific topic?
- What was unknown or what questions were the authors trying to answer?
- What was the authors' approach?
- Did the authors propose any hypotheses?
- What specialist terminology (jargon) do I need to define?

For the materials and methods

- What was the *general* approach?
- What *specific* methods were used?
- Am I familiar with these methods? (If not, acquire background information from secondary sources.)

For the results Look at each figure and read the figure caption to determine what kind of results were collected. Results can be descriptive or numerical.

Photographs, gel images, phylogenetic trees, maps, and flow charts typically show descriptive data. Tables may also contain descriptive data. Possible questions to ask about these kinds of visual aids include:

- What is the subject of the figure (or table)?
- Does the figure show a sequence of events? If so, what is that sequence?
- Are there any labeled organelles, structures, or marks? Why are they important?
- Does the picture show a relationship between form and function? If so, what is the relationship?
- Are there any noteworthy patterns? If so, what is the pattern?

Graphs always show quantitative (numerical) data. Look at each graph and identify the variables. By convention, the independent variable (the one the investigator manipulated) is plotted on the x-axis, and the dependent variable (the one that changes in response to the independent variable) is plotted on the y-axis. On bar graphs, one of the variables is typically categorical rather than quantitative.

- What was the relationship between the independent and dependent variables?

- If a hypothesis was tested, was there a difference between the controls and the treatment groups? If so, how were they different?

For the discussion The structure of the Discussion section is a triangle, narrow at the top and wide at the base. Information flows from specific to broad (just the opposite of the introduction). The first few paragraphs present the results along with the authors' interpretation. In the next part of the discussion, the results are compared with those in other research papers. The discussion often wraps up with the authors' main conclusions or how this work contributes to the body of knowledge on this topic.

- What were the main results?
- What do the results mean?
- What was the authors' most important conclusion?

Read selectively

With your list of questions in hand, you are now able to read the article with a specific goal: to find the answers to your questions. Tech-savvy readers who have downloaded PDF versions of journal articles may open them in Adobe Reader and enter key words from their questions to search the document for answers quickly and systematically. Whether you use the document search feature or not, write the answers in your own words. If the answers raise new questions, write them down, too, so that you can find those answers later.

Recite

Recite means to say out loud from memory. Whereas reading uses your eyes and writing uses your hands, reciting forces you to speak and listen. When you use multiple senses in a learning environment, you make more connections, which lead to deeper understanding and better recall. Go to a room where you won't be embarrassed to talk to yourself. Read each of your answers out loud. Listen to yourself speak. Are these words *you* would use or are they someone else's? Does one sentence follow logically from the previous one? Keep practicing until you can answer the questions in your own words without looking at your answer sheet. If you are having trouble with this step, have a conversation with your study group or instructor to clarify your understanding.

Review

Take a step back and review what you've learned. Were you able to find answers to all of your questions, both the original ones and the new ones? Are you unsure of any of your answers? The question and answer approach to reading makes it easier to articulate what it is that you don't understand. Get help and then review again. The review process gives you multiple opportunities to reflect on what you've learned, identify gaps in your understanding, and think about topics at a deeper level.

Share your knowledge

It's no secret that sharing what you've learned with others reinforces your own knowledge. That said, you probably had a specific reason for reading a journal article, and that reason determines *how* you share your knowledge. For example, you may be asked to analyze the results on an exam, or present the paper to your journal club or research group, or cite the paper in a lab report, research proposal, poster, or thesis to demonstrate that you are familiar with the most current knowledge on this topic. Other examples of how scientific knowledge is shared are given on p. 31–32.

Strategies for Reading Your Textbook

If a textbook is required for your course, you can be sure that your instructor expects you to read it. Keep up with your readings. Before each class, preread the text as described in "Survey the content" below. Go to class, take notes, and then actually *read* the assigned text, focusing on the topics emphasized in class. Reading for comprehension requires your full concentration. Turn off your cell phone and eliminate all other distractions for 30-40 minutes. Then take a short break. Repeat the process. You will find that you can accomplish much more in less time when you focus on one thing at a time.

The following strategy for reading your textbook is recommended by many university teaching and learning support centers (see, for example, resources posted by The Teaching and Learning Center at Bucknell University 2016; Counselling Services at the University of Victoria 2016; and links within these websites to other websites). The steps follow the **SQ3R method** (survey, question, read, recite, review), which was developed to help students learn and remember what they read. The process takes more time in the beginning, but it saves you time studying in the long run. The following steps work best with a chapter no longer than 25–30 pages.

Survey the content

This first step in the SQ3R method should not take a lot of time and should be done before class. Look over the assigned pages to get an overview of the content. Figure out the main topics by skimming the chapter title, the introduction, the end-of-chapter summary, and the check-your-understanding questions and problems. Then assess the level of detail by skimming the headings and subheadings as well as the pictures. Finally, scan the text for boldfaced terms, which are often vocabulary words that you are expected to know. Prereading the chapter before class allows you to spend less time writing and more time listening, because you already know what information is covered in your textbook.

Go to class and take notes

Some instructors provide PowerPoint slide decks for their lectures, or your textbook may come with a printed lecture notebook or a DVD with the figures. Bring these printouts to class to use when taking notes. Write down anything the instructor writes on the board. Write down anything associated with the words "This is important." Develop your own shorthand system for taking notes. If you missed something, insert a big question mark so that you can fill in the missing information later. Your notes form a framework for organizing information about the topic, and they help you identify the key concepts that were emphasized in lecture. With your notes as a reference, reading becomes an exercise in elaborating on details and making connections.

Formulate questions

Before you start reading, ask yourself, "What do I already know about this topic?" Make a list of key concepts that you remember. Keep this list handy so that you can correct any misconceptions after you finish reading.

Now review your notes and formulate questions about the topics. Reading engages the eyes, but thinking about the topics and writing down questions provides your brain with additional sensory input. The more senses you engage in learning, the better your memory recall. Here are some possible questions that will help you engage actively with the material.

- Why is [this topic] important?
- How is [topic 1] related to [topic 2]?
- How is [topic 1] different from [topic 2]?
- What experimental evidence led to our current understanding of this topic?
- Why was this approach to the problem taken? Would another also have worked?

Read selectively

Find the sections in your assigned reading that cover the topics emphasized in class. Fill in the gaps in your notes. Find the answers to your questions. As you do so, note any new questions that arise. Note any points of confusion. Define every word so that you become comfortable with the vocabulary. When symbols and formulas are involved, state in words what the terms mean. Interpret any graphs or other experimental data. Make diagrams and concept maps to help you see how topics are related.

Recite

After you're satisfied that you've found answers to your questions, say them out loud using your own words. The combination of speaking and listening engages two additional senses, enhancing your ability to process and remember the information. The acts of speaking and listening may also help you catch errors of logic and missed connections, especially when done in the presence of your study group or instructor. Repeat and refine your answers until you can recite them with confidence.

Review

Set aside a block of time at regular intervals to review your class notes and your reading notes. If you have trouble remembering all of the information, schedule your personal review sessions at more frequent intervals. Definitely try to answer any "Test Your Understanding" questions that come with your textbook. Work the problems without looking at the answers. When you get a wrong answer, try to pinpoint exactly where you went wrong. Articulating what you don't understand will help your instructor give you the kinds of cues that will allow you to figure out the answer for yourself.

Concept (mind) mapping

A **concept map** (also called a **mind map**) is a type of flow chart that links smaller concepts to a main concept. Mind maps are a way for readers to organize knowledge about a topic visually. Mind mapping is based on the premise that new knowledge must be integrated with existing knowledge before further learning is possible. Without this integration, new knowledge is quickly forgotten and misconceptions in existing knowledge will continue to persist (Novak and Cañas 2008). When used with an active reading strategy such as the SQ3R method described above, mind mapping is a powerful way to think about concepts more deeply and retain the information longer (University of Victoria Counselling Services, Reading and Concept Mapping Learning Module, 2016).

The strategy described here works best with a chapter or section of text no longer than 25–30 pages (Palmer-Stone 2001).

1. Take no more than 25 minutes to:
   - Read the chapter title, introduction, and summary (at the end of the chapter, if present).
   - Read the headings and subheadings.
   - Read the chapter title, introduction, summary, headings, and subheadings again.
   - Skim the topic sentence of each paragraph (usually the first or second sentence).
   - Skim italicized or boldfaced words.
2. Close your textbook. Take a full 30 minutes to:
   - Write down everything you can remember about what you read in the chapter (make a mind map). Each time you come to a dead end, use memory techniques such as associating ideas from your reading to lecture notes or other life experiences; visualizing pages, pictures, or graphs; staring out the window to daydream; and letting your mind go blank.
   - Figure out how all this material is related. Organize it according to what makes sense in your mind, not necessarily according to how it is organized in the textbook. Write down questions and possible contradictions to check on later.
3. Open your textbook. Fill in the blanks in your mind map with a different colored pencil.
4. Read the chapter again, this time normally. Make another mind map.
5. Review the material at regular intervals. If you can't construct your mind map in sufficient detail (i.e., you have forgotten much of what you read), then review more frequently.

How to Succeed in College

Success in college starts with having the right attitude. You may not be interested in every topic in every course, but it is important to look at the big picture. You *chose* to go to college because you expect that the knowledge and skills learned in college will help you achieve your short-term and long-term goals. *Meaningful* learning, however, requires a conscious decision to *want* to learn.

Assuming you are motivated to learn, how can you make the most of your study time? What counts is not how much time, but the quality of the time you spend studying. Take advantage of your school's resources to develop good time management and study skills in your first semester. Some strategies include putting due dates for all assignments in your calendar; planning ahead, especially when you have multiple exams or papers in the same week; scheduling study time in blocks of 30-40 minutes with breaks in between; reviewing class notes as soon as possible after class; reviewing material in short, frequent sessions rather than in one long, drawn-out sitting the day before an exam. Figure out where, when, and how you learn most effectively.

Finally, take advantage of all of your resources. Most biology textbooks come with instructions for using the book. Take the time to read those instructions, because they tell you how to use the various components —including online resources—most effectively. Join a study group (see below), attend all classes and recitations, and go to your instructor's office hours to seek help sooner rather than later. Keep in mind that learning takes time, effort, and repetition, and your willingness to learn makes all the difference in a successful academic experience.

Study Groups

If you have read the material several times, taken notes, and listened attentively in lecture, but still have questions, talk about the material with your classmates. Small study groups are one reason why students who choose to major in the sciences persist in the sciences, rather than switching to a non-science major (Light 2001).

What are some benefits of participating in small study groups? One benefit is the comfort level. You may be more likely to talk about problems when you are among your peers; after all, they are not the ones who assign your grade. Secondly, when a group is composed of peers with a similar knowledge base, group members speak the same language. Your instructor speaks a different language, because he or she has already struggled to master the material. When you communicate with your classmates, you verbalize your ideas at a level that is appropriate for your audience of peers. Finally, collaborative learning reflects the way scientists exchange information and share findings in the real world. A spirit of camaraderie develops when people work together toward a common goal. The prospect of learning difficult subject matter is no longer so daunting when you have support from a small group of like-minded individuals. The hard work may even be fun when there is good group chemistry.

Group study is not a substitute for studying alone, however. You must hold yourself accountable for reading the material, taking notes, and figuring out what you do not understand before you meet with your group. If you have not struggled to understand the material yourself, you are not in a position to help a classmate.

Plagiarism

Plagiarism is using someone else's ideas or work without acknowledging the source. Plagiarism is ethically wrong and demonstrates a lack of respect for members of your academic community (faculty and fellow students) and the scientific community in general. Many instructors are now using plagiarism checking services such as Turnitin® and SafeAssign™ by Blackboard to discourage *intentional* plagiarism, such as "borrowing" portions of another student's work, recycling lab reports from previous years, and buying papers on the Internet. Plagiarists who are caught can expect to receive at a minimum a failing grade on the assignment and close scrutiny in subsequent work. Plagiarism may also be cause for expulsion from school.

Many cases of plagiarism are *unintentional*, however, and stem from issues such as:

- Failure to understand what kind of information must be acknowledged
- Failure to reference the original material properly
- Failure to understand the subject matter clearly

Information that does not have to be acknowledged

General information that is obtained from sources such as news media, textbooks, and encyclopedias does not have to be acknowledged.

EXAMPLE: Most of the ATP in eukaryotic cells is produced in the mitochondria.

Information that is common knowledge for your audience does not have to be acknowledged. In an introductory course in cell and molecular biology, for example, students would be expected to know that ATP synthase is the enzyme that produces ATP through oxidative phosphorylation.

EXAMPLE: ATP is synthesized when protons flow down their electrochemical gradient through a channel in ATP synthase.

Information that has to be acknowledged

Information that falls into any of the following categories must be acknowledged:

- Information that is not widely known
- Controversial statements, opinions, or other people's conclusions
- Pictures or illustrations that you use but did not produce
- Statistics or formulas used in someone else's work
- Direct quotations

Paraphrasing the source text

Direct quotations are used in the humanities, but usually not in scientific papers. This idiosyncrasy of technical writing requires you to paraphrase the information in the source document. **Paraphrasing**—using your own words to express someone else's ideas—requires considerable thought and effort on your part. Not only do you have to have sufficient knowledge about the subject, you have to feel comfortable using the vocabulary. Read your textbook and other secondary sources, discuss the topic in your study group, or ask your instructor for clarification. A lot of groundwork has to be done before you can even begin to read a journal article, let alone paraphrase information it contains.

After you have sufficient background information on the topic, you are ready to tackle the content. Accept the fact that comprehension is an ongoing process in which you will read the source text, process the information you've read, read the text again, and process some more. When you are comfortable with the content, take notes on the important points, following the collective advice of Hofmann (2014), Lannon and Gurak (2011), McMillan (2012), Pechenik (2016), and other authorities on scientific writing:

- Don't take notes until you have read the source text at least twice.
- Don't look at the source text when you are taking notes.
- Use your own words and write in your own style.
- Retain key words.
- Don't use full sentences.
- Distinguish your own ideas and questions from those of the source text (e.g., "Me: Applies only to prokaryotes?").
- Use quotation marks to indicate exact or similar wording. Keep in mind that you will have to put the information into your own words if you use the information in your paper.
- Don't cite out of context. Preserve the author's original meaning.
- Give yourself permission to not understand everything. If it's important, get help.
- Fully document the source for later listing in the end references.

Faulty note-taking practices, particularly those that involve copying large portions of the original text, are likely to result in unintentional plagiarism. Beware of the pitfalls illustrated in Table 3.1. To avoid plagiarism, write in your own words and cite the source. For practice in identifying and avoiding plagiarism, take Frick's (2016) excellent online plagiarism tutorial. Read your institution's policies on academic responsibility, consult with professionals at your school's writing center, and ask your instructor for clarification when in doubt.

TABLE 3.1 Examples of plagiarism

Original Text
F_1 extends from the membrane, with the α and β subunits alternating around a central subunit γ. ATP synthesis occurs alternately in different β subunits, the cooperative tight binding of ADP + P_i at one catalytic site being coupled to ATP release at a second. The differences in binding affinities appear to be caused by rotation of the γ subunit in the center of the α3 β3 hexamer.

Plagiarized Text	Reason
According to Fillingame (1997), F_1 extends from the membrane, with the α and β subunits alternating around a central subunit γ. ATP synthesis occurs alternately in different β subunits, the cooperative tight binding of ADP + P_i at one catalytic site being coupled to ATP release at a second. The differences in binding affinities appear to be caused by rotation of the γ subunit in the center of the α3 β3 hexamer.	The author's actual words were used without quotation marks or indenting the citation. Because direct quotations are not used in scientific papers, it is imperative that you paraphrase. Using the original text is plagiarism even when the source is cited.
F_1 consists of α and β subunits alternating around a central subunit γ. In the β subunits, tight binding of ADP + P_i occurs at one catalytic site and ATP is released at a second. The different binding affinities may be caused by rotation of the γ subunit in the center (Fillingame 1997).	The basic sentence structure of the original text was maintained. A few words were omitted or changed, but the text is still highly similar to the original.
ATP synthase consists of a transmembrane protein (F_o), a central shaft (γ), and an F_1 head made up of α and β subunits. As protons enter F_o, the shaft rotates, changing the conformation of the β subunits, allowing ADP and P_i to bind and be released as ATP.	The text was paraphrased, but the source of the information was not cited.

Source: From Fillingame RH. 1997. Coupling H^+ transport and ATP synthesis in F_1F_o-ATP synthases: glimpses of interacting parts in a dynamic molecular machine. *The Journal of Experimental Biology* accessed 2017 Jan 19: 200: 217–224. http://jeb.biologists.org/content/200/2/217.full.pdf+html

The Benefits of Learning to Write Scientific Papers

Why is it valuable to learn how to write scientific papers? First, scientific writing is a systematic approach to describing a problem. By writing what you know (and what you do not know) about the problem, it is often possible to identify gaps in your own knowledge.

Second, the scientific method is a logical approach to answering questions. It involves coming up with a tentative solution, gathering information to become more knowledgeable about the topic, evaluating the reliability of the information, testing and analyzing the data, and arriving at a reasonable conclusion. This approach can be applied to many situations in your life, from deciding which graduate school to apply to, to choosing your next cell phone or your first new car.

Third, when you learn to write lab reports, you are investing in your future. Publications in the sciences are affirmation from your colleagues that your work has merit; you have been accepted into the community of experts in your field. Even if your career path is not in the sciences, scientific writing is very logical and organized, characteristics appreciated by busy people everywhere.

Credibility and Reputation

The credibility and reputation of scientists are established primarily by their ability to communicate effectively through their written reports. Poorly written papers, regardless of the importance of the content, may not get published if the reviewers do not understand what the writer intended to say.

You should think about your reputation even as a student. When you write your laboratory reports in an accepted, concise, and accurate manner, your instructor knows that you are serious about your work. Your instructor appreciates not only the time and effort you spent learning the subject matter, but also your willingness to write according to the standards of the profession.

Model Papers

Before writing your first laboratory report, look at articles published in biology journals such as *American Journal of Botany*, *Ecology*, *The EMBO Journal*, *Journal of Biological Chemistry*, *Journal of Molecular Biology*, and *Marine Biology*. Download or photocopy one or two journal articles that interest you so you can refer to them for format questions.

Almost all journals devote one page or more to "Instructions to Authors," in which specific information is conveyed regarding length of the manuscript, general format, figures, conventions, references, and so on. Skim this section to get an idea of what journal editors expect from scientists who wish to have their work published.

Because most beginning biology students find journal articles hard to read, sample student laboratory reports are given in Chapter 6. Read the comments in the margins as you peruse the reports to familiarize yourself with the basics of scientific paper format and content, as well as purpose, audience, and tone.

Chapter 4

Step-by-Step Instructions for Preparing a Laboratory Report or Scientific Paper

In order to prepare a well-written laboratory report according to accepted conventions, the following skills are required:

- A solid command of the English language
- An understanding of the scientific method
- An understanding of scientific concepts and terminology
- Advanced word processing skills
- Knowledge of computer graphing software
- The ability to read and evaluate journal articles
- The ability to search the primary literature efficiently
- The ability to evaluate the reliability of Internet sources

If you are a first- or second-year college student, it is unlikely that you possess all of these skills when you are asked to write your first laboratory report. Don't worry. The instructions in this chapter will guide you through the steps involved in preparing the first draft of a laboratory report. Revision is addressed in the next chapter, and the Appendices will help you with word processing and graphing tasks.

Timetable

Preparing a laboratory report or scientific paper is hard work. It will take much more time than you expect. Writing the first draft is only the first step. You must also allow time for editing and proofreading (revision). If you work on your paper in stages, the final product will be much better than if you try to do everything at the last minute.

TABLE 4.1 Timetable for writing your laboratory report

Time Frame	Activity	Rationale
Day 1	Complete laboratory exercise.	It's fun. Besides, you need data to write about.
Days 2–3	Write first draft of laboratory report.	The lab is still fresh in your mind. You also need time to complete the subsequent tasks before the due date.
Day 4	Proofread and revise first draft (hard copy).	Always take a break after writing the first draft and before revising it. This "distance" gives you objectivity to read your paper critically.
Day 5	Give first draft to a peer reviewer for feedback, *if your instructor permits it.*	Your peer reviewer is a sounding board for your writing. He/she will give you feedback on whether what you intended to write actually comes across to the reader. You may wish to alert your peer reviewer to concerns you have about your paper (see "Get Feedback" in Chapter 5).
	Arrange to meet with your peer reviewer after he/she has had time to review your paper ("writing conference").	An informal discussion is useful for providing immediate exchange of ideas and concerns.
Day 6	Peer reviewer reviews laboratory report.	The peer reviewer should review the paper according to two sets of criteria. One is the conventions of scientific writing as described in "Scientific Paper Format" in Chapter 3, and the other is the set of questions in "Get Feedback" in Chapter 5.
	Hold writing conference during which the reviewer returns the first draft to the writer.	An informal discussion between the writer and the reviewer is useful to give the writer an opportunity to explain what he/she intended to accomplish, and for the reviewer to provide feedback.
Days 6–7	Revise laboratory report.	Based on your discussion with your reviewer, revise as necessary. Remember that you do not have to accept all of the reviewer's suggestions.
Day 8	Hand in both first draft and revised draft to instructor.	Your instructor wants to know what you've learned (we never stop learning either!).

The timetable outlined in Table 4.1 breaks the writing process down into stages, based on a one-week time frame. You can adjust the time frame according to your own deadlines.

Format your report correctly

Although content is important, the appearance of your paper is what makes the first impression on the reader. Before submitting papers electronically, print out and proofread the hard copy. You will be surprised at how formatting and other kinds of errors jump out at you when you read on paper instead of on screen. When you turn in assignments on paper, make sure the pages are in order and the print is legible. Subconsciously or not, the reader/evaluator is going to associate a sloppy paper with sloppy science. You cannot afford that kind of reputation. In order for your work to be taken seriously, your paper has to have a professional appearance.

Scientific journals specify the format in their "Instructions to Authors" section. If your instructor has not given you specific instructions, the layout specified in Table 4.2 will give your paper a professional look.

Consult the sample "good" student laboratory report in Chapter 6 for an overview of the style and layout. An electronic file called "Biology Lab Report Template," available at <http://sites.sinauer.com/Knisely5E> is formatted according to the guidelines of Table 4.2 and provides prompts that help you get started writing in scientific paper format. For details on how to format documents in Microsoft Word, see the "Formatting Documents" section in Appendix 1.

Computer savvy

Know your computer and your word processing software. Most of the tasks you will encounter in writing your laboratory report are described in Appendix 1, "Word Processing in Microsoft Word" and Appendix 2, "Making Graphs in Microsoft Excel." If there is a task that is not covered in these appendices, write it down and ask an expert later. If you run into a major problem that prevents you from using your computer, you should have a backup plan in place (access to another computer).

Always back up your files somewhere other than your computer's hard drive. Options may include a USB flash drive (also called a jump drive or thumb drive), an external hard drive, or online. Online options include cloud services such as Google Drive, saving your files to your organization's server, or emailing files to yourself. See the section "Backing up your files" on p. 193 in Appendix 1 for more information.

Save your file frequently while writing your paper by clicking on the Quick Access Toolbar. You can also adjust the settings for automatically saving your file. Windows users, click **File | Options**. Mac users, click **Word | Preferences | Output and Sharing**. From there, the common sequence is **Save | Save AutoRecover information every __ minutes**.

TABLE 4.2 Instructions to authors of laboratory reports

Feature	Layout
Paper	8½″ x 11″ (or DIN A4) white bond
Margins	1.25″ left and right; 1″ top and bottom
Font size	12 pt (points to the inch)
Typeface	Times Roman or another *serif* font. A serif is a small stroke that embellishes the character at the top and bottom. The serifs create a strong horizontal emphasis, which helps the eye scan lines of text more easily.
Symbols	Use word processing software. Do not write symbols in by hand.
Pagination	Arabic number, top right on each page except the first
Justification	Align left/ragged right or Full/even edges
Spacing	Double
New paragraph	Indent 0.5″
Title page (optional)	Title, authors (your name first, lab partner second), class, and date
Headings	Align headings for Abstract, Introduction, Materials and Methods, Results, Discussion, and References on left margin or center them. Use consistent format for capitalization. Do not start each section on a new page unless it works out that way coincidentally. Keep section heading and body together on the same page.
Subheadings	Use sparingly and maintain consistent format.
Tables and figures	Incorporate into text as close as possible after the paragraph where they are first mentioned. Use descriptive titles, sequential numbering, proper position above or below visual. May be attached on separate pages at end of document, but must still have proper caption. Keep table/figure and its caption together on the same page.
Sketches	Hand-drawn in pencil or ink. Other specifications as in "Tables and figures" above.
References	Citation-Sequence System: Make a numbered list in order of citation. Name-Year System: List references in alphabetical order by the first author's last name. Use a hanging indent (all lines but the first indented) to separate individual references. Both systems: Use accepted punctuation and format.
Assembly	Place pages in order, staple top left.

Install antivirus software on your computer and always check flash drives for viruses before you use them. Beware of files attached to email messages. Do not open attachments unless you are sure they come from a reliable source.

Store flash drives with their caps on to keep dust out. Protect them from excess humidity, heat, and cold. Only remove a flash drive from a computer after you eject it and the message "Safe to Remove Hardware" is displayed.

If you must eat and drink near a computer, keep beverages and crumbs away from the hard drive and keyboard.

Getting Started

Set aside 1 hour to begin writing the laboratory report as soon as possible after doing the laboratory exercise. Turn off your phone and get off social networking sites. Writing lab reports requires your full concentration. What matters is the quality, not the quantity, of time you spend on your assignments. Promise yourself a reward for time well spent.

Reread the laboratory exercise

You cannot begin to write a paper without a sense of purpose. What were the objectives of your experiment or study? What questions are you supposed to answer? Take notes on the laboratory exercise to prevent problems with plagiarism when you write your laboratory report.

Organization

If your instructor provided a rubric or other instructions for organizing your lab report, follow the instructions exactly. Otherwise use the standard IMRD format, as described in Chapter 3.

Audience

Scientific papers are written for scientists. Similarly, laboratory reports should be written for an audience of fellow student-biologists, who have a knowledge base similar to your own. When deciding how much background information to include, assume that your audience knows only what you learned in class. Use scientific terminology, but define any terms or acronyms known only to experts (**jargon**).

Write for an audience of fellow scientists, not students in a classroom situation. Note the difference between the original text and the revision in the following examples:

FAULTY: The experiments performed by the students dealt with how different wavelengths of light affect seed germination.

REVISION: The purpose of the experiment was to determine how different wavelengths of light affect seed germination.

FAULTY: The purpose of this experiment is to become acquainted with new lab techniques such as protein analysis, serial dilutions, and use of the spectrophotometer.

REVISION: The purpose of this experiment was to use the biuret assay to determine protein concentration in egg white.

Writing style

Laboratory reports are formal written assignments. Avoid slang and contractions and choose words that reflect the serious nature of scientific study. Readers of scientific papers trust the scientific method and are confident that the facts speak for themselves. For this reason, write objectively—that is, do not make judgments. When making a statement that may not be obvious to the audience, always back it up by citing an authoritative source or by providing experimental evidence. Because the focus is on the science, not the scientist, passive voice is used more frequently (especially in the Materials and Methods section) than in other kinds of writing. Use active voice in the other sections, however, because it makes sentences shorter and more dynamic.

Past and present tense have specific connotations in scientific papers. Authors use *present tense* to make *general statements* that the scientific community agrees are valid. Statements that are generally valid include explanations of phenomena based on experimental results that have been replicated by many scientists. Therefore, use present tense in the Introduction and Discussion sections when describing information accepted by the scientific community, and cite the source of any information that is not common knowledge for your audience. On the other hand, authors use *past tense* to make statements about *their own work*. For this reason, use past tense in the Materials and Methods and Results sections, and whenever you are describing work that you personally carried out.

Start with the Materials and Methods Section

The order in which you write the different sections is not the order in which they appear in the finished laboratory report. The rationale for this plan will become obvious as you read on. The Materials and Methods section requires the least amount of thought, because you are primarily restating the procedure in your own words.

Tense

When you write the Materials and Methods section, describe the procedure in *past*, not present, tense because (1) these are completed actions and (2) you are describing your own work. Do *not* copy the format of your laboratory exercise, in which the instructions may be arranged in a numbered list and the imperative (command) form of verbs may be used for clarity.

Voice

There are two grammatical voices in writing: active and passive. In active voice, the subject *performs* the action. In passive voice, the subject *receives* the action. Passive voice is preferred in the Materials and Methods section because the subject that receives the action is more important than who performed it. The logic is that anyone with the appropriate training should be able to perform the action. Consider the following examples:

ACTIVE VOICE: I peeled and homogenized the turnips.

PASSIVE VOICE: The turnips were peeled and homogenized.

The sentence written in active voice is more natural and dynamic, but it shifts the emphasis from the subject, "the turnips" to "I." Passive voice places the emphasis on the turnips, where it belongs. Because sentences written in passive voice tend to be longer and less direct than those written in active voice, try to use active voice when the performer (you) is not the subject of the sentence.

Level of detail

A well-written Materials and Methods section will *provide enough detail to allow someone with appropriate training to repeat the procedure.* For example, for a **molecular biology** procedure, include essential details such as the concentration and pH of solutions, reaction and incubation times, volume, temperature, wavelength (set on a spectrophotometer), centrifugation speed, dependent and independent variables, and control and treatment groups. On the other hand, *do not describe routine lab procedures* such as:

- How to calculate molarity or use $C_1V_1 = C_2V_2$ to make solutions.
- Taring a balance before use.
- Using a vortex mixer to ensure that solutions are well mixed.
- Describing how to zero (blank) a spectrophotometer before measuring the absorbance of the samples.
- Explaining what type of serological pipette or micropipettor is appropriate for a particular volume.
- Designating the type of flasks or beakers to use.
- Specifying the duration of the entire study ("In our two-week experiment, …").

For a **field experiment**, however, time *is* important. When observing or collecting plants and animals in nature, be sure to include in the Materials and Methods section time of day, month, and year as appropriate; sampling frequency; location and dimensions of the study site; sample size; and statistical analyses. Depending on the focus of your lab report, it may also be prudent to describe the geology, vegetation, climate, natural history, and other characteristics of the study site that could influence the results.

Here are some guidelines for the level of detail to include in the Materials and Methods section.

Not enough information Include all relevant information needed to repeat the experiment.

FAULTY: In this lab, we mixed varying amounts of BSA stock solution with varying amounts of TBS using a vortex mixer. We used a spectrophotometer to measure absorbance of the 4 BSA samples, and then we determined the concentration of 4 dilutions of egg white from the standard curve.

EXPLANATION: This procedure does not give the reader enough information to repeat the experiment, because essential details like *what concentrations of BSA* were used to construct the standard curve, *what dilutions of egg white* were tested, and the *wavelength* set on the spectrophotometer have been left out.

REVISION: Bovine serum albumin (BSA) solutions (2, 3, 5, 10 mg/mL) were prepared in tris-buffered saline (TBS). The egg white sample was serially diluted 1/5, 1/15, 1/60, and 1/300 with TBS. The absorbance of all samples was measured at 550 nm using a Spec 20 spectrophotometer.

The following are examples of **too much information**.

Do not list materials and methods separately The wording of the section heading makes it tempting to separate the content into two parts. In fact, materials should not be listed separately unless the strain of bacteria, vector (plasmid), growth media, or chemicals were obtained from a special or noncommercial source. It will be obvious to the reader what materials are required on reading the methods.

Describe the solutions, not the containers

FAULTY: Eight clean beakers were labeled with the following concentrations of hydrogen peroxide and those solutions were created and placed in the appropriate beaker: 0, 0.1, 0.2, 0.5, 0.8, 1.0, 5.0, and 10.0.

EXPLANATION: Using clean, suitable containers to store solutions is common practice in the laboratory. Putting labels on labware is also a routine procedure. An essential detail missing from this sentence is the units.

REVISION: The following hydrogen peroxide solutions were prepared: 0, 0.1, 0.2, 0.4, 0.8, 1.0, 5.0, and 10.0%.

Specify the concentrations, not the procedure for making solutions

FAULTY: To make the dilution, a micropipette was used to release 45, 90, 135, and 180 µL of bovine serum albumin (BSA) into four different test tubes. To complete the dilution, 255, 210, 165, and 120 µL of TBS was added, respectively.

EXPLANATION: With appropriate instruction, making dilutions of stock solutions becomes a routine procedure. In the above example, you should assume that your readers can make the solution using the appropriate measuring instruments *as long as you specify the final concentration.*

REVISION: The following concentrations of BSA were prepared for the Bradford assay: 300, 600, 900, and 1200 µL/mL.

Include only essential procedures and write concisely

FAULTY: The test tubes were carried over to the spectrophotometer and the wavelength was set to 595 nm (nanometer). The spectrophotometer was zeroed using the blank. Each of the remaining 8 samples in the test tubes were individually placed into the empty spec tube, which was then placed in the spectrophotometer where the absorbance was determined.

EXPLANATION: The only detail important enough to mention is the wavelength.

REVISION: The absorbance of each sample was measured with a Spectronic 20 spectrophotometer at 595 nm.

Avoid giving "previews" of your data analysis

FAULTY: A graph was plotted with Absorbance on the *y*-axis and Protein concentration on the *x*-axis. An equation was found to fit the line, then the unknown protein absorbances that fell on the graph were plugged into the equation, and a concentration was found.

EXPLANATION: Making graphs is something that you do when you analyze your raw data, but it is not part of the experimental procedure. How and why you chose to plot the data will become obvious to the reader in the Results section, where you display graphs, tables, and other visuals and describe the noteworthy findings.

REVISION: Delete this entire passage.

Cite published sources If you are paraphrasing a published laboratory exercise, it is necessary to cite the source (see "Documenting Sources" on pp. 89–101). Unpublished laboratory exercises are not usually cited; ask your instructor to be sure.

Do the Results Section Next

The Results section is a *summary* of the key findings of your experiment. This section has two components:

- Visuals, such as tables and figures
- A body, or text, in which you describe the results shown in the visuals

When you work on the Results section, you will complete the following tasks, which are often done concurrently, not necessarily sequentially:

- **Analyze the raw data.** Raw data are all the observations and measurements that you recorded in your lab notebook. It is your job as the author to analyze all these data and process the information for the reader. **Do not simply transfer raw data into your lab report** (your instructor may ask you to attach pages from your lab notebook as an appendix, however). Instead, summarize the data by eliminating aberrant results (because you realized that you made a mistake in obtaining these results), averaging replicates, using statistical methods to see possible trends, and/or selecting representative pictures (for example, micrographs or gel images). The goal of data analysis in general is to try to figure out what the data show. More specifically, you compare the results to the predictions that were based on the hypotheses you proposed when designing your experiment. When the results match the predictions, then the hypotheses are supported. Conversely, when the results are unexpected, further research may be required.
- **Organize summarized data in tables or figures.** When you organize summarized data in a table or plot numerical values on a graph, you may be able to see trends that were not apparent before. Effective visuals are more powerful than words alone and they provide strong support for your arguments. See the "Preparing visuals" section on pp. 60–72.
- **Decide in which order to present the tables and figures.** The sequence should be logical, so that the first visual provides a basis for the next or so that the reader can easily follow your line of reasoning.
- **Describe each visual in turn and refer to it in parentheses.** Describe the most important thing you want the reader to notice about the visual. Refer to the visual by number in parentheses at

TABLE 4.3 Types of graphs and their purpose

Graph	Purpose	Example
Histogram	To show the distribution of a quantitative variable.	Distribution of grades on an exam. *Y*-axis shows number of students; *x*-axis shows numerical score on the exam.
Scatterplot	To show the relationship between two quantitative variables measured on the same individuals. Look for an overall pattern and for deviations from that pattern. If the points lie close to a straight line, a linear trendline may be superimposed on the scatter graph. The correlation, *r*, indicates the strength of the linear relationship.	Relationship between shell length and mass. If we are just looking for a pattern, it doesn't matter which variable is plotted on which axis. If we suspect that mass depends on length, plot mass on the *y*-axis and length on the *x*-axis. Look at the form, direction, and strength of the relationship.
Line graph	To show the relationship between two quantitative variables. One variable may be dependent on the other. The variable that is being manipulated is called the independent or explanatory variable. The variable that changes in response to the independent variable is called the dependent or response variable. By convention, the independent variable is plotted on the *x*-axis and the dependent variable is plotted on the *y*-axis. Error bars may be included to show variability.	Relationship between enzyme activity and temperature. Because temperature is the variable that is being manipulated, it is plotted on the *x*-axis. Because enzyme activity is the response being measured, it is plotted on the *y*-axis.

the end of the first sentence in which you describe it. See "Writing the body of the Results section" on pp. 75–80.

Preparing visuals

The most common visuals in scientific writing are tables and figures. A **table** is defined by Webster's dictionary as "a systematic arrangement of data usually in rows and columns for ready reference." A **figure** is any visual that is not a table. Thus, line graphs (also called XY graphs), bar graphs, pie graphs (also called pie charts), drawings, gel photos, X-ray images, and microscope images are all called *figures* in scientific papers.

The type of visual you use depends on the objectives of your study or experiment and the nature of the data. Use a **table** when

TABLE 4.3 *(continued)*

Graph	Purpose	Example
Scatterplot with regression line	To predict the value of *y* for a given value of *x* or vice versa. The response variable must be dependent on the explanatory variable and the relationship must be linear. The regression line takes the form $y = mx + b$, where *m* is the slope and *b* is the *y*-intercept.	Standard curve for a protein assay. Protein concentrations of a standard such as BSA are plotted on the *x*-axis. Absorbance (measured by a spectrophotometer) for each concentration is plotted on the *y*-axis. A regression line is fitted to the data. To predict the protein concentration of a sample (*x*), measure its absorbance (*y*) and solve the regression equation for *x*.
Bar graph	To show the distribution of a categorical (non-quantitative) variable.	Effect of different treatments on plant height. One axis shows the treatment category and the other shows the numerical response.
Pie graph	To show the distribution of a categorical (non-quantitative) variable in relation to the whole. All categories must be accounted for so that the pie wedges total 100%.	Composition of insects in a backyard survey. Each wedge represents the percentage of an order of insects. Orders with low representation may be combined into an "Other" wedge to complete the pie.

- The exact numbers are more important than the trend.
- Statistics such as sample size, standard error, and P-values are used to support your conclusions.
- Arranging categorical variables and other non-quantitative information makes it easier to interpret the results.

Use a **graph** to show relationships between or among variables. The type of graph that can be used is often dictated by the nature of the variables—quantitative or categorical. **Categorical variables** are groups or categories that have no units of measurement (treatment groups, age groups, habitat, etc.). Bar graphs and pie graphs are commonly used to display results involving categorical variables. **Quantitative variables,** on the other hand, have numerical values with units. XY graphs and scatter graphs (also called scatterplots) display relationships between quantitative variables. Some of the graphs frequently encountered in the field of biology are summarized in Table 4.3 and described individually in the following sections.

Do not feel that you have to have visuals in your lab report. If you can state the results in a sentence, then no visual is needed (see Example 1 in the "Organizing Your Data" section).

Tables

Tables are used to display large quantities of numbers and other information that would be tedious to read in prose. Arrange the categories vertically, rather than horizontally, as this arrangement is easier for the reader to follow (see, for example, Table 1 in Figure 4.1). List the items in a logical order (e.g., sequential, alphabetical, or increasing or decreasing value). Use sentence case for the headings. Include the units in each col-

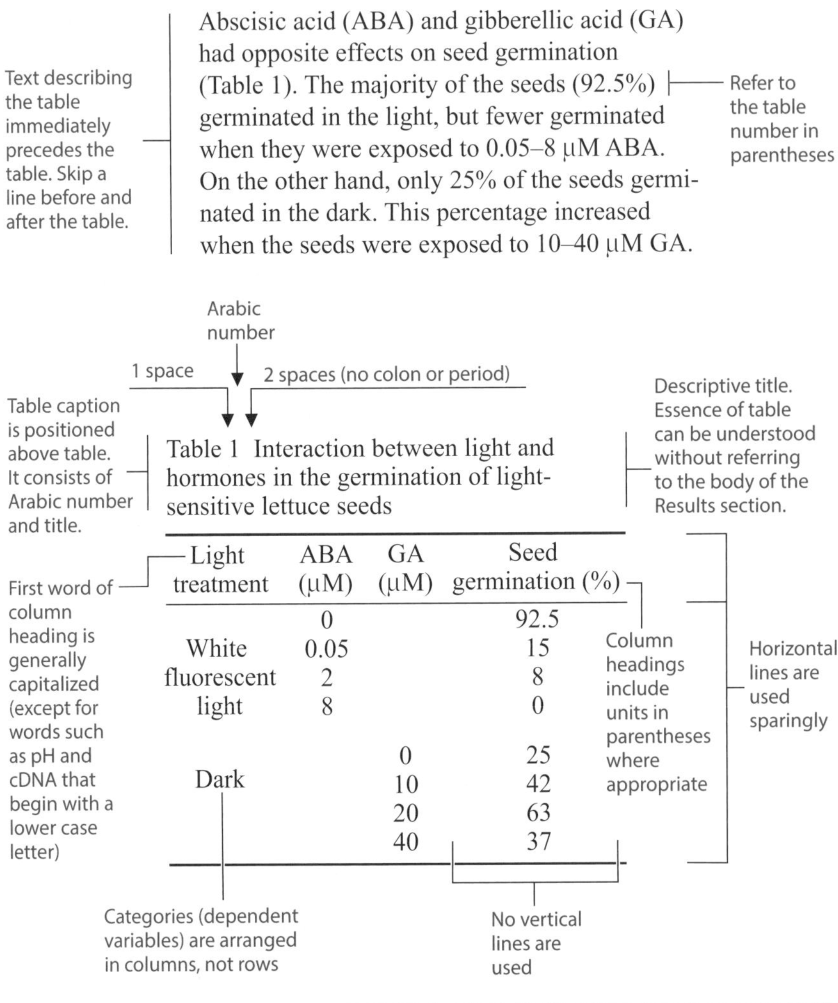

Abscisic acid (ABA) and gibberellic acid (GA) had opposite effects on seed germination (Table 1). The majority of the seeds (92.5%) germinated in the light, but fewer germinated when they were exposed to 0.05–8 μM ABA. On the other hand, only 25% of the seeds germinated in the dark. This percentage increased when the seeds were exposed to 10–40 μM GA.

Table 1 Interaction between light and hormones in the germination of light-sensitive lettuce seeds

Light treatment	ABA (μM)	GA (μM)	Seed germination (%)
White fluorescent light	0		92.5
	0.05		15
	2		8
	8		0
Dark		0	25
		10	42
		20	63
		40	37

Figure 4.1 Excerpt from a Results section showing a properly formatted table preceded by the text that describes the data in the table.

umn heading to save yourself the trouble of writing the units after each number entry in the table.

By convention, tables in scientific papers do not have vertical lines to separate the columns, and horizontal lines are used only to separate the table caption from the column headings, the headings from the data, and the data from any footnotes. The tables in this book are formatted in this style.

Table captions Give each table a caption that includes a number and a title. Center the caption or align it on the left margin *above* the table. Use Arabic numbers, and number the tables consecutively in the order they are discussed in the text. Notice that in this book, the table and figure numbers are preceded by the chapter number. This system helps orient the reader in long manuscripts, but is not necessary in short papers like your laboratory report.

Table titles From the table title alone, the reader should be able to understand the essence of the table without having to refer to the body (text) of the Results section. For simple tables, it may suffice to use a precise noun phrase rather than a full sentence for the title. For more complex tables, one or more full sentences may be required. Either way, English grammar rules apply:

- Do not capitalize common nouns (*general* classes of people, places, or things) unless they begin the phrase or sentence.
- Capitalize proper nouns (names of *specific* people, places, or things).
- Do not capitalize words that start with a lower case letter (for example, pH, mRNA, or cDNA), even if they begin a sentence.

Some examples of faulty and preferred titles are shown below.

FAULTY: Table 1 The Relationship Between Light and Hormones in the Germination of Light-Sensitive Lettuce Seeds

EXPLANATION: Use sentence case. Do not capitalize common nouns unless they start the sentence.

FAULTY: Table 1 Table of interaction between light and hormones in the germination of light-sensitive lettuce seeds

EXPLANATION: Do not start a title with a description of the visual.

FAULTY: Table 1 Seed germination data

EXPLANATION: Do not write vague and undescriptive titles.

REVISION: Table 1 Interaction between light and hormones in the germination of light-sensitive lettuce seeds

A table is always positioned *after* the text in which you refer to it (see Figure 4.1). Refer to the table number in parentheses at the end of the first sentence in which you describe the table contents. That way the reader can refer to the table as you describe what you consider to be important.

In your laboratory report, it is not necessary to include a table when you already have a graph that shows the same data. Make *either* a table *or* a graph—not both—to present a given data set.

Tables can be constructed in either Microsoft Word (see "Tables" in Appendix 1) or Microsoft Excel (see Formatting the Spreadsheet—Tables in Appendix 2).

Table preparation checklist

- ☐ Categories arranged in columns, not rows
- ☐ Column headings include units (where appropriate)
- ☐ Format correct (minimal lines)
- ☐ Table title descriptive
- ☐ Table title in sentence case
- ☐ Table caption positioned above the table

Line graphs (XY graphs) and scatterplots

Line graphs display a relationship between **two or more quantitative variables**. To avoid confusion between the terms *line graph* and Excel's *line chart*, I will use *XY graph* when referring to line graphs. Excel's line chart is a special case of XY graph in which the independent variable is a constant interval such as time. If the values for the independent variable are not equally spaced, then Excel's **scatter chart** should be used to plot the data.

Figure 4.2 Different XY graph formats showing relationships between variables. (A) A scatterplot displays numerical observations with the purpose of determining whether there is a relationship between shell mass and shell length. (B) A scatterplot with a straight line added shows that there is a strong linear relationship between the two variables. (C) The relationship between the independent and dependent variables in each treatment group is hard to see when the points are not connected. (D) The relationship is much easier to see when the points are connected with straight or smoothed lines. (E) Error bars show variability about the mean. (F) A least-squares regression line and its equation are used to predict one variable when the other is known. Linear regression lines are used only in specific situations when the mathematical relationship between the two variables is clearly established. ▶

What we hope to learn from the data in part determines the format of the XY graph (Figure 4.2). For example, in **observational studies**, scientists observe individuals and measure variables that they are interested in. Quite often, the purpose of an observational study is to look for a pat-

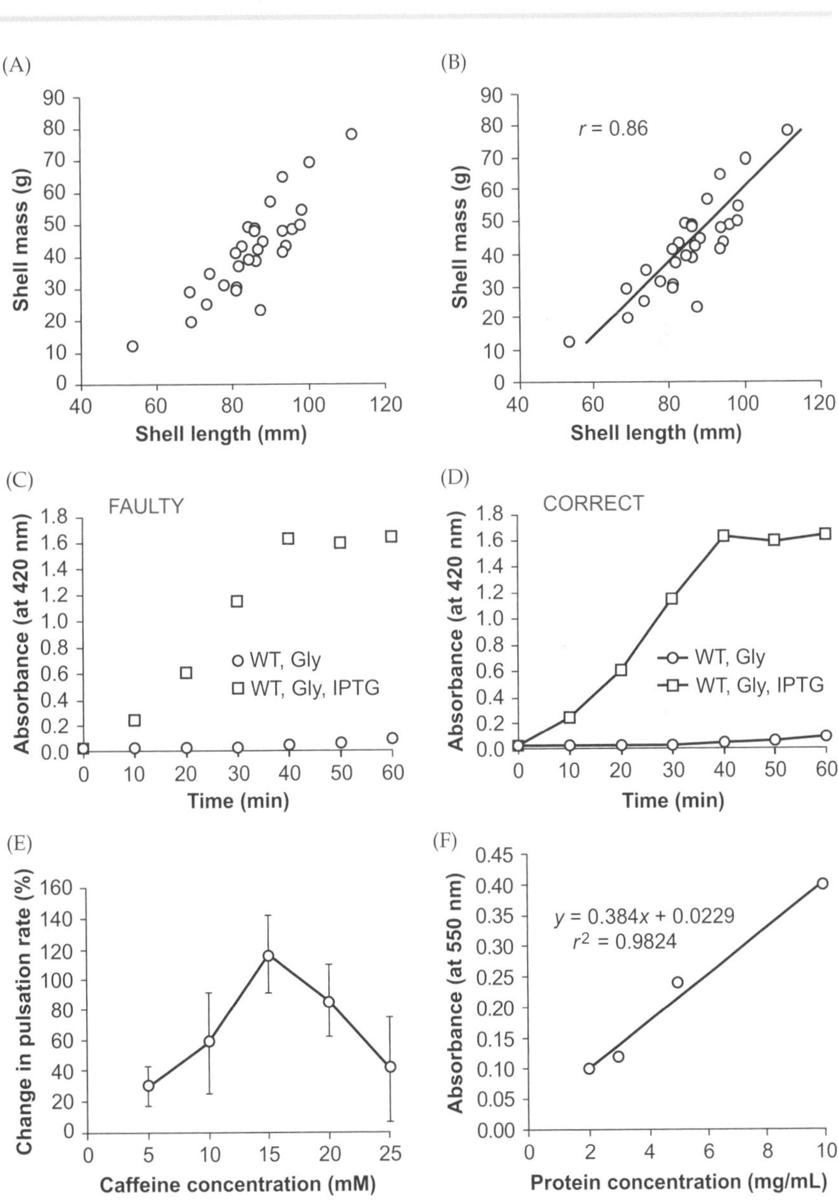

tern in nature. Patterns may be easier to spot when the numerical data are plotted as a **scatterplot**, one kind of XY graph in which the individual data points are not connected (see Figure 4.2A). By convention, if one of the variables explains or influences the other, then this so-called explanatory or **independent variable** is plotted on the x-axis. The variable that shows the response, also called the **dependent variable**, is plotted on the y-axis. In some observational studies, there may not be a causative relationship between the two variables, in which case it doesn't matter which variable is plotted on which axis.

The first step in describing a pattern on a scatterplot is to look at the form and direction of the data. The **form** may be linear, curved, clustered, or random; because many relationships in nature are linear, exponential, or logarithmic, keep an eye out for these kinds of forms. The **direction** indicates whether the relationship between the variables is positive (large values for y correspond to large values for x and vice versa) or negative (large values for y correspond to small values for x and vice versa), or if there is no change in y with x or vice versa. Once you've described the form and direction of the scatterplot, try to assess the **strength** of the relationship. How closely do the points follow the form? A lot of scatter and the presence of outliers indicate a weak relationship.

Our eyes are pretty good at recognizing when the data fall on a straight line, but we need a more objective way to assess the strength of the relationship. One such indicator is called **correlation** (r), whose rather complex formula produces values between –1 and 1. Correlation values near 0 indicate a weak linear relationship, with the strength of the relationship increasing as r approaches –1 (when the direction is negative) or 1 (when the direction is positive). When the data in an observational study show a strong linear relationship, scientists may superimpose a straight line on the scatterplot and display r as a measure of the strength of the relationship (see Figure 4.2B).

In observational studies, scientists measure a variable of interest without trying to influence the response. On the other hand, in **experiments**, scientists impose a treatment on individuals and then observe how the treatment affects their responses. The purpose of an experiment, therefore, is to determine the effect of one variable (the explanatory or independent variable) on another (the response or dependent variable). By convention, the independent variable (the one the scientist manipulates) is plotted on the x-axis and the dependent variable (the one that changes in response to the independent variable) is plotted on the y-axis.

On a scatterplot, the individual data points are not connected, because the purpose of the graph is to determine the form, direction, and strength of the relationship between the variables. In contrast, in an experiment, scientists want to know **how the imposed treatments affect the response**. To make it easier to see this effect, **the data points are connected by**

straight or smoothed lines. Lines avoid confusion particularly when there is more than one data set on a graph (compare Figures 4.2C and D). Never show the lines without the experimental data, however.

Data points displayed on graphs are typically a summary of the raw data, with each point representing the mean value calculated by averaging many replicates. **To show variability in the measured values** (especially when the data are distributed normally about the mean), authors may include **error bars** on their graphs (see Figure 4.2E). An explanation of what the error bars represent—standard deviations or standard errors of the mean—should be given in the figure title along with the number of observations.

Finally, **standard curves** represent a special case of XY graph, whose purpose is to predict one variable when the other is known. First the data points are plotted as a scatterplot, and then a least-squares regression line (best-fit line) is fitted to the points (see Figure 4.2F). The square of the correlation, r^2, describes how well the regression line fits the data. The closer the r^2 value is to 1, the better the fit. The better the fit, the closer the predicted value will be to the true value of the unknown variable.

Figure captions Figures are always numbered and titled *beneath* the visual (Figure 4.3). The captions may be centered or placed flush on the left margin of the report. Arabic numbers are used, and the figures are numbered consecutively in the order they are discussed in the text.

Figure titles From the figure title alone, the reader should be able to understand the essence of the figure without having to refer to the body (text) of the Results section. For simple figures, it may suffice to use a precise noun phrase rather than a full sentence for the title. For more complex figures, one or more full sentences may be required. Either way, English grammar rules apply: Do not capitalize common nouns (*general* classes of people, places, or things) unless they begin the phrase or sentence. Capitalize proper nouns (names of *specific* people, places, or things). Do not capitalize words that start with a lower case letter (for example, pH, mRNA, or cDNA), even if they begin a sentence. Some examples of faulty and preferred titles for the graph in Figure 4.3 are shown here.

FAULTY: Figure 1 The Effect of Population Density on the Development of Male Gametophytes

EXPLANATION: Use sentence case. Do not capitalize common nouns unless they start the sentence.

FAULTY: Figure 1 Percentage of male gametophytes vs. population density

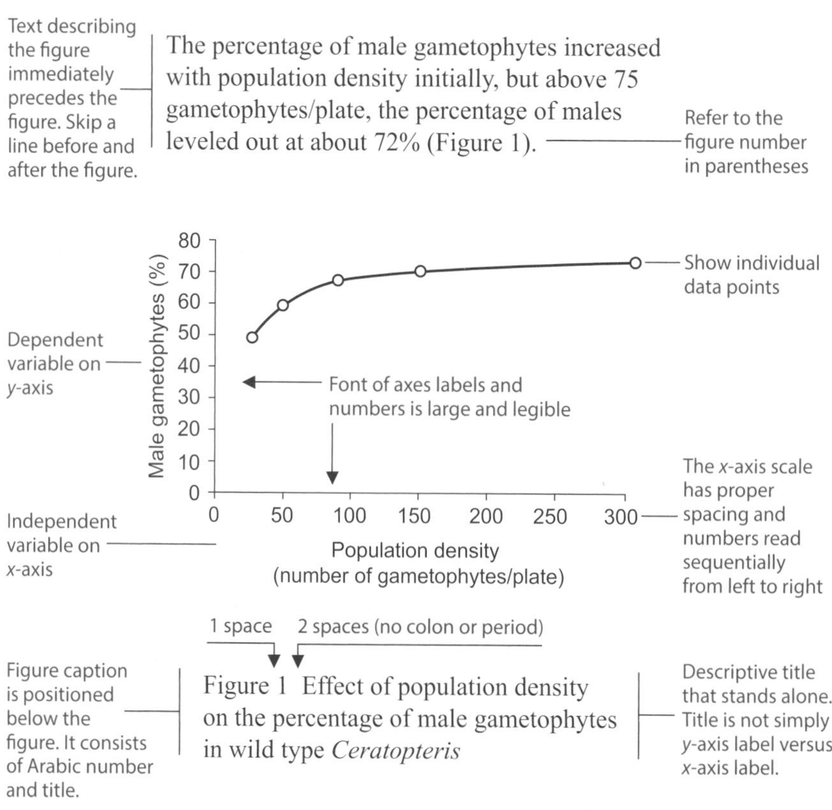
Text describing the figure immediately precedes the figure. Skip a line before and after the figure.
The percentage of male gametophytes increased with population density initially, but above 75 gametophytes/plate, the percentage of males leveled out at about 72% (Figure 1).
Refer to the figure number in parentheses
80
70
60
50
40
30
20
10
0
Male gametophytes (%)
Show individual data points
Dependent variable on y-axis
Font of axes labels and numbers is large and legible
0
50
100
150
200
250
300
Population density
(number of gametophytes/plate)
Independent variable on x-axis
The x-axis scale has proper spacing and numbers read sequentially from left to right
1 space
2 spaces (no colon or period)
Figure caption is positioned below the figure. It consists of Arabic number and title.
Figure 1 Effect of population density on the percentage of male gametophytes in wild type Ceratopteris
Descriptive title that stands alone. Title is not simply y-axis label versus x-axis label.

Figure 4.3 Excerpt from a Results section showing a properly formatted figure with one line (data set); the text that describes the figure precedes it.

EXPLANATION: Do not restate the *y*-axis label versus the *x*-axis label as the figure title.

FAULTY: Figure 1 shows the effect of population density on the percentage of male gametophytes in wild type *Ceratopteris*

EXPLANATION: Separate the figure number and the title.

FAULTY: Figure 1 Line graph of the effect of population density on the percentage of male gametophytes in wild type *Ceratopteris*

EXPLANATION: Do not start a title with a description of the visual.

FAULTY:	Figure 1 Averaged class data for C-fern experiment
EXPLANATION:	Do not write vague and undescriptive titles.
REVISION:	Figure 1 Effect of population density on the percentage of male gametophytes in wild type *Ceratopteris*

More than one data set When there is more than one data set (line) on the figure, you have three options:

- Add a brief label (no border, no arrows) next to each line.
- Use a different symbol for each line and label the symbols in a legend (as in Figure 4.4). Place the legend without a border within the axes of the graph. This is the easiest option if you are using Excel to plot your data.

The concentration of cells in both cultures remained low for several hours (Figure 2). At about 4 hr, the concentration of the ABC strain increased rapidly and then leveled off after about 9 hr. On the other hand, it took longer for the concentration of strain XYZ to increase, but at 12 hr, the concentration was much higher than that of strain ABC.

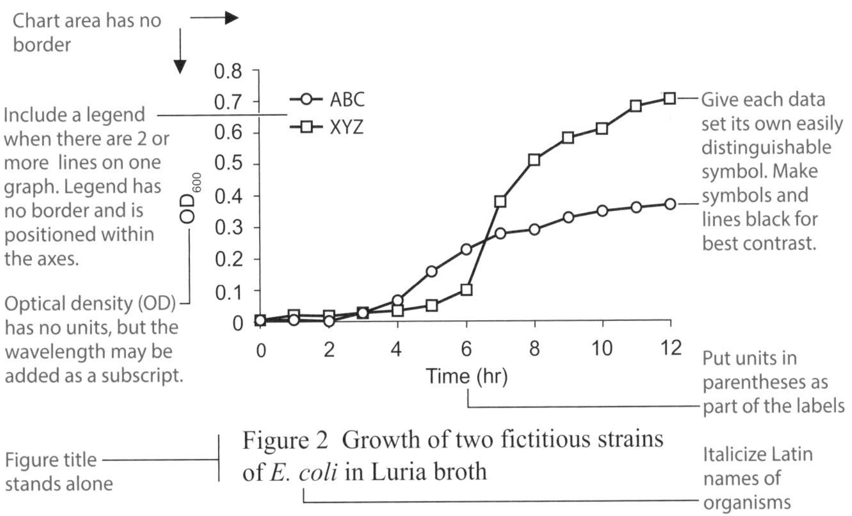

Figure 2 Growth of two fictitious strains of *E. coli* in Luria broth

---

Figure 4.4 Excerpt from a Results section showing a properly formatted figure with two sets of data. A legend is needed to distinguish the two lines. The text that describes the figure precedes it.

- If the first two options make the figure look cluttered, identify the symbols in the figure caption.

All three formats are acceptable in scientific papers as long as you use them consistently.

Black and white or color Figures in your laboratory report should be prepared according to the guidelines specified by the Council of Science Editors (CSE Manual 2014). If this graph will be included in a written assignment such as a lab report, then make all lines, numbers, and symbols black for best contrast on a white background. If this graph will be used on a poster or in a PowerPoint presentation, colored lines make the graph visually appealing. Black is preferred for written documents because, when they are copied or printed on a black and white printer, there is no ambiguity about black on white. Colored lines turn out to be various shades of gray, which may make it hard for the reader to distinguish the numbers on the axes and the different data sets.

Although you may wish to plot a rough draft of your graphs by hand, you should learn how to use computer plotting software to make graphs. Microsoft Excel is a good plotting program for novices (see Appendix 2) because it is readily available and fairly easy to use. The time you invest now in learning to plot data on the computer will be invaluable in your upper-level courses and later in your career.

Bar graphs

A bar graph allows you to compare individual sets of data when **one of the variables is categorical** (not quantitative)—this is the main difference between XY graphs and bar graphs. Bar graphs are more flexible than pie charts because any number of categories can be compared; the percentages do not have to total 100%. Error bars may be centered at the top of each data bar to show variability in the measured data. When the data bars are black, only half error bars are used.

Consider an experiment in which you want to compare the final height of the same species of plant treated with four different nutrient solutions. The nutrient solution is the non-numerical, categorical variable; the height is the response variable. The data bars are arranged vertically in Figure 4.5, because the category labels are short.

Figure 4.6, on the other hand, is an example of a bar graph with horizontal bars. This arrangement is more convenient to accommodate the long categorical labels that describe the rats' maternal diet and diet after weaning. The dependent variable is the time the rats spent searching for the hidden escape platform in the memory retention portion of the Morris water maze test.

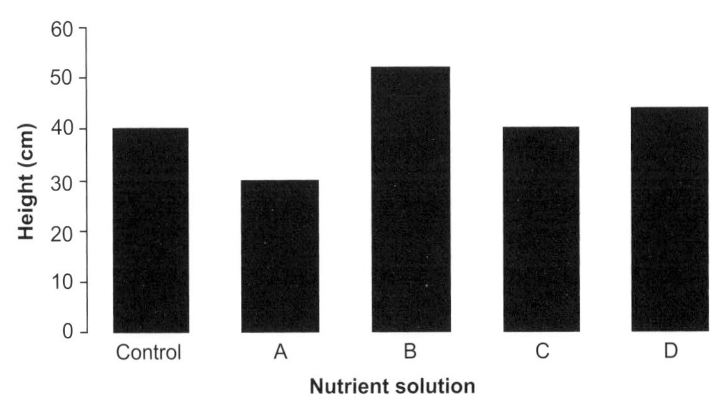

Figure 4.5 Final height of corn plants after 4-week treatment with different nutrient solutions. This figure is an example of a column graph.

The bars should be placed sequentially, but if there is no particular order, then put the control treatment bar far left in column graphs or at the top in horizontal bar graphs. Order the experimental treatment bars from shortest to longest (or vice versa) to facilitate comparison among the different conditions. The baseline does not have to be visible, but all the bars must be aligned as if there were a baseline.

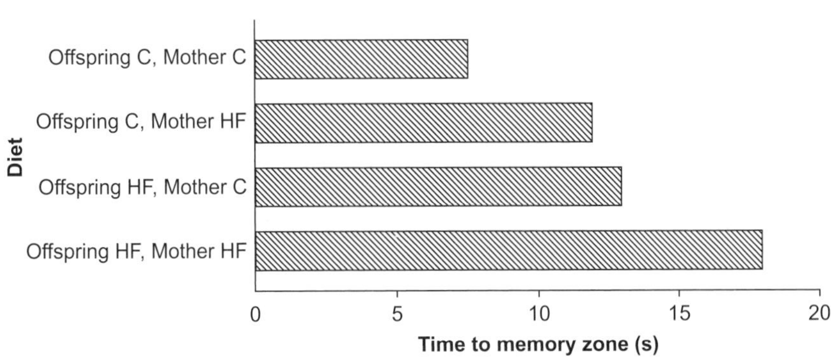

Figure 4.6 The time male rats spent swimming to the location of a submerged platform in the probe test portion of the Morris water maze. The study was conducted to determine the effect of maternal and post-weaning diet on memory retention in rats. HF = high fat diet, C = control diet. Data kindly provided by Professor Kathleen Page, Bucknell University. This figure is an example of a horizontal bar graph with long category labels and pattern bars.

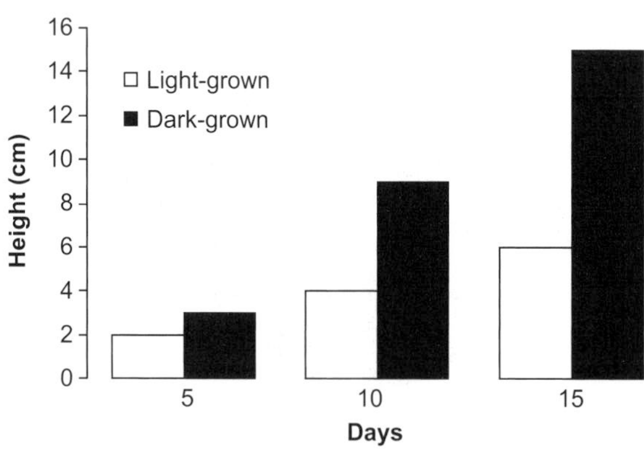

Figure 4.7 Difference in height of groups of light-grown and dark-grown bean seedlings at 5, 10, and 15 days after planting. This figure is an example of a clustered bar graph. Each bar in the cluster must be easily distinguishable from its neighbor.

The bars should always be wider than the spaces between them. In a graph with clustered bars, make sure each bar has sufficient contrast so that it can be distinguished from its neighbor (Figure 4.7). Instructions for plotting bar graphs in Excel 2013 and Excel for Mac 2016 are given in Appendix 2.

Pie graphs

A pie graph is used to show data as a percentage of the total data. For example, if you were doing a survey of insects found in your backyard, a pie graph would be effective in showing the percentage of each kind of insect out of all the insects sampled (Figure 4.8). There should be between two and eight segments in the pie. Place the largest segment in the right-hand quadrant with the segments decreasing in size clockwise. Combine small segments under the heading "Other." Position labels and percentages horizontally outside of the segments for easy reference. Instructions for plotting pie charts in Excel 2013 and Excel for Mac 2016 are given in Appendix 2.

Figure preparation checklist

- ☐ Right type of graph
- ☐ Format correct (symbols, lines, legend, axis scale, outside tick marks, no gridlines, no border)
- ☐ Figure title descriptive
- ☐ Figure title in sentence case
- ☐ Figure caption positioned below the figure

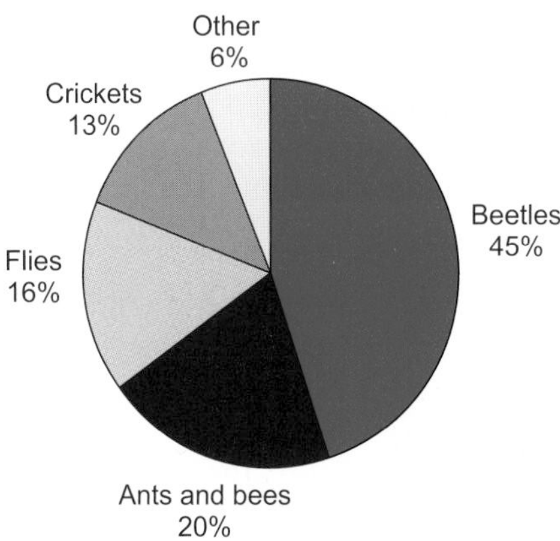

Figure 4.8 Composition of insects in backyard survey. Pie graphs are used to show data as a percentage of the total data.

Organizing your data

Reread the laboratory exercise to see if your instructor has provided specific guidelines on the kinds of visuals to include in the Results section. If you have to make the decision on your own, ask yourself the following questions:

- Can I state the results in one sentence? If so, then **no visual** is needed.
- Are the numbers themselves more important than the trend shown by the numbers? If so, then use a **table**.
- Is the trend more important than the numbers themselves? If so, use a **graph**.
  - Are both variables quantitative? If so, then use an **XY graph**.
  - Is one of the variables categorical (not quantitative)? If so, then use a **bar graph**.
- Are the results descriptive rather than quantitative? If so, use **photos and images**.

The following examples demonstrate that there may be more than one good way to organize the data. Some visuals may be more appropriate than others, and in some cases, no visual may be the best alternative.

EXAMPLE 1: *Brassica* seeds were placed on filter paper saturated with pH 1, 2, 3, or 4 buffered solutions. The positive control was filter paper saturated with water. After 2 days, 100% of the seeds in the positive control germinated. No seeds germinated in any of the buffered solutions.

POSSIBLE SOLUTION: *No visual* is needed because the results can be summarized in one sentence: "After 2 days, 100% of the seeds that imbibed water germinated, but none of the seeds that were treated with buffered solutions pH 1, 2, 3, or 4 germinated."

EXAMPLE 2: Light-sensitive lettuce seeds placed on filter paper saturated with water were exposed to the same fluence of white fluorescent, red, far-red, green, and blue light treatments as well as darkness, and the percentage that germinated was determined 30 hours later.

POSSIBLE SOLUTION: Since some of the data are categorical rather than quantitative (colors rather than wavelengths of light), either a *table* or a *bar graph* (Figure 4.9) works well to display the results.

INAPPROPRIATE: An *XY graph* is not appropriate because both variables are not quantitative. We only know the colors of light, not the exact wavelengths. *Text only* is **not** appropriate because listing the seed germination percentages in a sentence is tedious to read and hard to comprehend.

EXAMPLE 3: The activity of an enzyme (catalase) was monitored at nine different temperatures in order to determine the optimal temperature for maximum activity.

POSSIBLE SOLUTION: If your emphasis is on the actual numbers rather than the trend, then display the results in a *table*. If the trend is more important than the numbers, use an *XY graph and connect the points* with smoothed or straight lines (Figure 4.10).

(A) Table 1 Effect of light treatment on percentage of light-sensitive lettuce seeds germinated after 30 hr

Light treatment	Seed germination (%)
Red	76
White	65
Blue	44
Far-red	38
Green	37
Dark	30

(B)

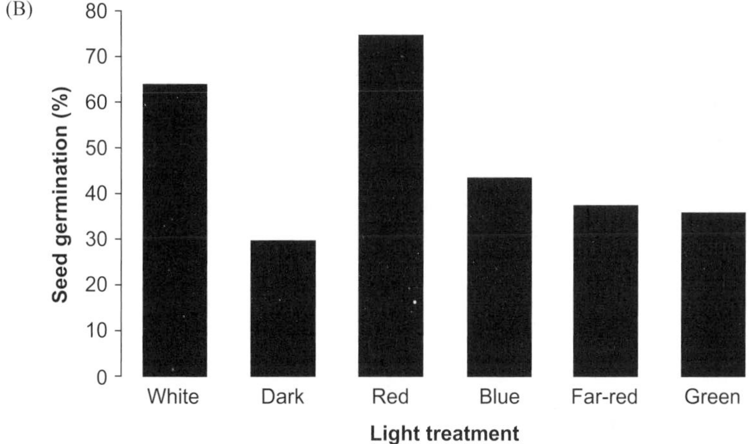

Figure 1 Effect of light treatment on percentage of light-sensitive lettuce seeds germinated after 30 hr

Figure 4.9 Example 2 data summarized in (A) a table and (B) a bar graph. The positive and negative controls are placed to the left, and, if there is no particular order to the categories (colors in this example), arrange the bars in order of the longest to the shortest (or vice versa).

Writing the body of the Results section

Now that your data are displayed visually in graphs or tables, it's time to tell your audience what you consider to be important. The suggestions below will help you get started.

(A)

Table 1 Effect of temperature on catalase activity

Temperature (°C)	Catalase activity (units of product formed · sec^{-1})
4	0.039
15	0.073
23	0.077
30	0.096
37	0.082
50	0.040
60	0.007
70	0
100	0

(B)

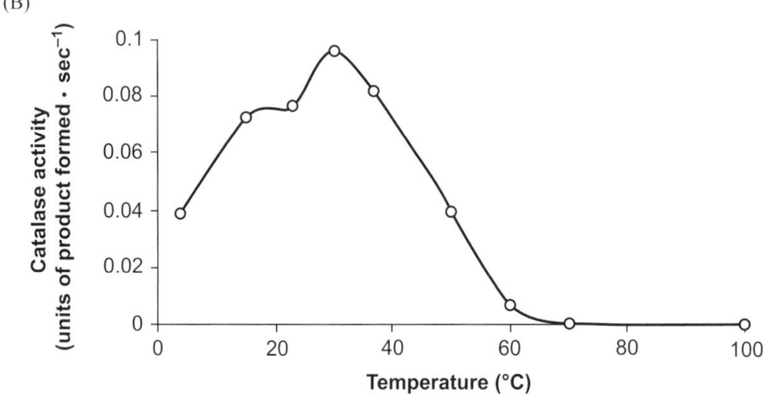

Figure 1 Effect of temperature on catalase activity

Figure 4.10 Example 3 data summarized in (A) a table and (B) an XY graph. When data are summarized in a table, the emphasis is placed on the numbers rather than on the trend. For the same data, an XY graph is more effective than a table in showing the trend.

Use words that describe relationships that the reader can readily see in the figure How does the dependent variable change with the independent variable? Is there a recognizable mathematical relationship (for example, linear, logarithmic, or exponential) or does the trend require a more detailed description?

For example, Figure 4.4 shows an excerpt from a Results section. In this experiment, the growth of two fictitious strains of bacteria was monitored over time. The dependent variable, optical density (OD_{600}), is a measure of the concentration of cells in the culture. Consider how well each of the following sentences describes the graph.

VAGUE: The concentration of cells increased over time in both strains of *E. coli* (Figure 2).

BETTER: The concentration of cells remained low at first and then increased over time in both strains of *E. coli* (Figure 2).

BETTER YET: In both strains of *E. coli*, the concentration of cells was low for the first few hours, increased rapidly, and then leveled off after about 12 hr (Figure 2).

How much detail you use to describe the shape of the curve depends on how you plan to explain the results in the Discussion section. Pointing out key features in the Results section prepares the reader for what is to come in the discussion. The discussion is where you would correlate the changes in slope of your graph with the lag, log, and stationary phases of bacterial growth.

When there is more than one data set or category on a figure, make comparisons

TEDIOUS: In the ABC strain, concentration was low for the first few hours, increased rapidly after 4 hr, and then leveled off (Figure 2). In the XYZ strain, concentration was low for the first few hours, increased rapidly after 6 hr, and then leveled off.

BETTER: In both strains of *E. coli*, the concentration of cells was low for the first few hours, increased rapidly, and then leveled off (Figure 2). The concentration in the ABC culture began to increase after 4 hr and reached a maximum after about 12 hours. In contrast, the concentration of XYZ cells started to increase later, but reached a higher level after 12 hours.

Don't explain or interpret the results

FAULTY: Both strains of *E. coli* displayed a lag phase, followed by a log growth phase in which the cells divided rapidly, and then a stationary phase in which the growth rate was counterbalanced by the death rate (Figure 2).

EXPLANATION: The terms *lag phase, log growth phase,* and *stationary phase* are not marked on the graph, nor can the reader tell from the graph what the cells are doing. Save interpretations like this for the Discussion section.

Refer to the figure that shows the result you are describing

FAULTY: In both strains of *E. coli,* concentration was low for the first few hours, increased rapidly, and then leveled off.

EXPLANATION: Tell the reader where to find the data. Reference the figure in parentheses at the end of the sentence.

FIGURE AS SUBJECT: Figure 2 shows that in both strains of *E. coli,* concentration was low for the first few hours, increased rapidly, and then leveled off.

PREFERRED: In both strains of *E. coli,* concentration was low for the first few hours, increased rapidly, and then leveled off (Figure 2).

It is not incorrect to make the figure reference the subject of the sentence. However, notice how this style places the emphasis on the figure, instead of the results. The preferred style emphasizes the results, which is what the reader is looking for.

When you need more than one sentence to describe the results in a visual, refer to the figure just once in that paragraph, preferably at the beginning. That way, your readers will know right away where to look for the data, and they will assume you are describing the same figure unless you tell them otherwise.

Some XY graphs are intended to be used as tools, not to show relationships For example, standard curves are specific types of XY graphs that show absorbance as a function of concentration. Their purpose is not to show that absorbance is proportional to concentration (Beer's Law already establishes that fact), but to allow the concentration of an unknown to be calculated from its measured absorbance. When writing about standard curves, don't describe the trend. Instead, describe how the graph was used to calculate the unknown parameter of interest.

Make every sentence meaningful

FAULTY: Concentration changed over time in both strains of *E. coli* (Figure 2).

EXPLANATION: This sentence does not say *how* concentration changed over time.

FAULTY: The results in Figure 2 show the averaged data for the whole class.

EXPLANATION: This sentence does not describe an actual result. Describe statistical methods in the Materials and Methods section. Alternatively, state that these are averaged data, and include the number of trials in the figure caption.

FAULTY: After the results were obtained, a graph was made with time on the *x*-axis and concentration on the *y*-axis, as shown in Figure 2.

EXPLANATION: This sentence states the obvious and should be deleted. Describe results, not axis labels.

Eliminate unnecessary introductory phrases. This includes phrases such as

- It was found that…
- The results showed that…
- It could be determined that…

Get to the point! State the important results in clear, concise terms.

REVISION: Delete the introductory phrase and begin the sentence with an actual result.

Use past tense Whenever you are referring specifically to your own results, use past tense. In scientific papers, present tense is reserved for statements accepted by the scientific community as fact. At this point, your results are not yet considered "fact."

Equations

Equations are technically part of the text and should *not* be referred to as figures. Equations are set off from the rest of the text on a separate line. If you have several equations and need to refer to them unambiguously in the body of the Results (or another) section, number each equation sequentially and place the number in parentheses on the right margin. For example:

$$\text{Absorbance} = -\log T \qquad (1)$$

In Microsoft Word, to center an equation and right-align the equation number, insert a center tab stop and a right tab stop as shown.

When you are asked to show your calculations, use words to describe your calculation procedure, as in the following example. Listing equations without guiding the reader through the process is like including figures without pointing out the important results. Make your writing reader-friendly!

> Protein concentration of the unknown sample was determined using the equation of the biuret standard curve (Figure 1). The measured absorbance value was substituted for y, and the equation was solved for x (the protein concentration):
>
> $$y = 0.0417x$$
>
> $$0.225 = 0.0417x$$
>
> $$5.40 = x$$
>
> Thus, the protein concentration of the sample was 5.40 mg/mL.

When you present a sequence of calculations like this, align the = symbol in each line.

Type equations using MS Word's Equation Editor, accessed by clicking **Equation** on the **Insert** tab. Type the first equation into the box. Press **Enter**. Repeat the process for each equation in the group. To align the group of equations on the equal sign, select all of the equations, right-click, and select **Align at =**. This method does not work if each line has a right-aligned equation number. In that case, the equations have to be aligned manually.

Make Connections

Now that the "meat" of your report is done, it's time to describe how your work fits into the existing body of knowledge. These connections are made in the Discussion and Introduction sections. Ideally, there will be a one-to-one correspondence between the objectives stated in the Introduction and the interpretation and explanation of the results in the Discussion.

Write the discussion

The Discussion section gives you the opportunity to **interpret your results, relate them to published findings, and explain why they are important**. The structure of the discussion is specific to broad, as illustrated by the following triangle.

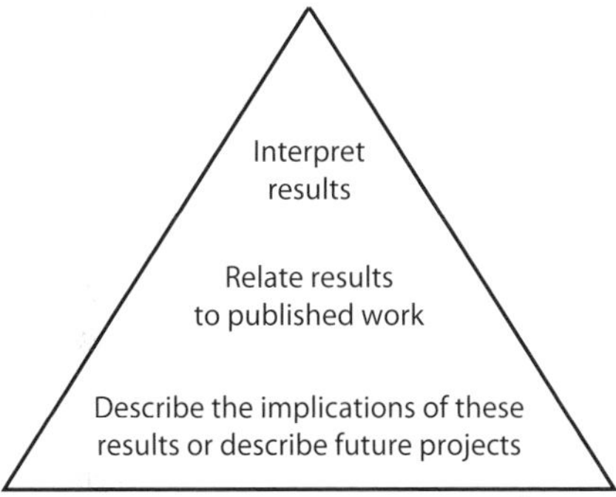

A springboard How do you start a Discussion section? There is no specific formula, but here are some suggestions.

- Summarize your most important result and then explain what it means.
- State whether your hypothesis was negated or supported and then provide the evidence.
- Reference a published journal article and state whether or not your results supported those findings.

If, initially, you don't know what your results mean, consult your resources. Does your lab handout give you any clues? Did your instructor talk about this topic in lecture? Don't forget to check the index of your textbook. When you think you have an explanation, make sure you use your own words and cite your sources. No matter how you start, by the time you finish your discussion, you should have interpreted your results in the context of your objectives and the work of other scientists.

When results defy explanation Especially in introductory biology labs, the results may not always work out the way we expect. If your results defy explanation, consider these possible reasons:

- Human error, including failure to follow the procedure, failure to use the equipment properly, failure to prepare solutions correctly, variability when multiple lab partners measure the same thing, and simple arithmetic errors. If you suspect that human error negatively affected your results, then say so, but don't waste a lot of words in doing so.
- Numerical values were entered incorrectly in the computer plotting program. If possible, correct the errors and repeat the analysis.
- Sample size was too small. If possible, collect more samples.
- Variability was too great to draw any conclusions. Consider redesigning your experiment.

If you can rule out these possibilities, discuss your results with your teaching assistant or instructor. Having a discussion with a knowledgeable individual may help you better understand the concepts, even if your results didn't turn out the way you expected.

Results never prove hypotheses When explaining your results, never use the word *prove*. Instead, use words and phrases like *provide evidence for, support, indicate, demonstrate,* or *strongly suggest*. The reason for this choice of words lies in the logic behind the scientific method. If our results match our predictions, then there is evidence that our hypotheses are correct. When many scientists get the same results independently, then the support for a given hypothesis grows. Scientists are reluctant to use the word "prove," because there is always a chance that a future study may provide conflicting evidence.

> **FAULTY:** These results prove that catalase was denatured at temperatures above 60 °C.
>
> **REVISION:** These results strongly suggest that catalase was denatured at temperatures above 60 °C.

Build your case Writing a Discussion section is an exercise in persuasive writing. You want to convince readers that your conclusions are valid. To do so requires substantiating statements with experimental evidence and referencing the work of other researchers.

To illustrate this approach, let's look at a study on the effect of human activities on the biodiversity of gastropods in rocky intertidal areas in southern California (Roy *et al.* 2003)*. The authors suggested that trampling,

---

*Roy K, Collins AG, Becker BJ, Begovic E, Engle JM. 2003. Anthropogenic impacts and historical decline in body size of rocky intertidal gastropods in southern California. Ecology Letters 6: 205–211.

shell collecting, and harvesting of these mollusks for food or bait caused a decrease in body size over time. To test this hypothesis, the authors examined museum shell collections dating back to the late 1800s, and they also

Chronology of the Discussion section	Rationale
Authors state that adults collected prior to 1960 were larger than those surveyed more recently.	This statement follows up on an observation made in the Introduction, that humans tend to harvest or collect large specimens.
Data are presented in Figure 2: Museum specimens collected before 1960 were significantly larger than those collected between 1961 and 1980 and those surveyed in the field.	Results support the first statement.
Data are presented in Table 1: There was a decrease in median size and in the size of the largest individuals.	Results provide additional support for the first statement.
Authors propose a counterargument: Another factor, such as climate change, could be responsible for the decrease in size.	This statement shows that the authors considered other explanations for the results.
Authors suggest a way to test the counterargument: If the decrease in body size is due to human activities, then study sites in which human activity is prohibited should have larger sized mollusks.	The authors propose a testable hypothesis to differentiate the effect of human activities and non-human activities.
Data are presented in Figure 3: Protected sites had larger individuals.	Results support the human activities hypothesis.
Authors compare their findings to those of other researchers: Similar results were found for other species in California (3 papers cited). Similar results were found for species in other parts of the world (5 papers cited).	Shows that the results of the current study support previous work (an indication that the results are trustworthy). Shows that the authors are aware of other studies in this field; they acknowledge the work of the other researchers.
Authors describe future research: Investigate all possible reasons why large gastropods are disappearing from the southern CA coast; investigate ways to reverse negative human impact.	The authors recognize that there may be more than one reason why body size is declining. Future work may test alternative hypotheses.

surveyed rocky intertidal sites where some of the museum specimens had been collected. Some of the survey sites were protected from human activities; at those sites, collection of invertebrates was prohibited.

In summary, this discussion is well organized, and the authors provide compelling evidence for their conclusions. Potentially controversial statements are supported with experimental results. Counterarguments were considered and refuted. References to other published papers enhance the credibility of this study.

Tense When *describing* your own results, use *past* tense. However, when you use scientific fact to *explain* your results, use *present* tense.

PAST TENSE: The initial velocity of the reaction *was* zero at temperatures between 60 °C and 90 °C (Figure 1).

EXPLANATION: Past tense signifies that you are describing your own results.

PRESENT TENSE: At high temperatures, there *is* no enzymatic activity because the enzymes *are* denatured.

EXPLANATION: Present tense signifies that these statements are generally valid and considered to be scientific fact.

Compare your results with those in the literature Do your results agree with those in published papers? If so, then your work supports the existing body of knowledge. If not, could a different method account for a conflicting result? If warranted, discuss possible reasons why your results did not turn out as expected.

Describe future work You might propose hypotheses and experiments that build on your results.

Write the introduction

After having written drafts of the Materials and Methods, Results, and Discussion sections, you should be intimately familiar with the procedure, the results, and what the results mean. Now you are in a position to put your study or experiment into perspective. What was already known about the topic? Were there any unanswered questions? Why did you carry out this work?

The structure of the introduction is broad to specific, just the opposite of that of the discussion.

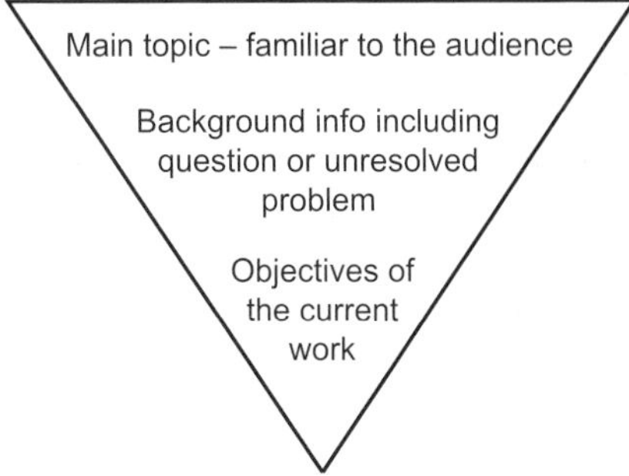

Organization The introduction consists of two main parts:

- Background information from the literature, and
- Objectives of the current work.

The opening sentence of the Introduction section is usually a **general observation or result** familiar to readers in that discipline. The author then quickly **narrows down the topic** and **provides background information** from the literature. The author then sets the stage for the current study by stating **any unanswered questions** or inconsistencies with previous work. Finally, in the last paragraph of the Introduction section, the author states the **objectives of the current work**. Specific hypotheses may be included if the study lends itself to hypothesis testing.

Start at the end Deciding how much background information to include is a daunting task. To make this task slightly less daunting, **write your introduction in reverse order**. In other words, **start with the objectives** and then gradually fill in the *minimum* amount of information your reader would need to understand why you chose those particular objectives. You may already have directly or indirectly addressed your objectives in the Discussion section. In that case, transcribe those sentences into the Introduction section. Alternatively, get inspiration from looking at the variables on the figures in the Results section. XY graphs typically show the effect of the *x*-axis variable on the *y*-axis variable. Photographs are used to show relationships between form and function. Flow diagrams illustrate processes. Gel images show bands that represent the size and amount of a particular nucleic acid or protein. Maps show distributions. Phylogenetic trees show evolutionary relationships. By studying the relationships, patterns, and structures shown in the figures and tables, you should be able to identify the specific objectives of your study.

Now work backward from your objectives. **What concepts would the reader need to be reminded of for these objectives (or specific hypotheses) to make sense?** For example, if you set out to determine how temperature affects the rate of a particular enzyme-substrate reaction, the reader would need to know how enzyme and substrate molecules interact and how temperature affects the motion of molecules. **How much detail should you include about these broad concepts?** Less is better. In other words, provide just enough detail to prepare your readers for what is to come in the Results and Discussion sections. Include only the details that are directly relevant to your study. Do not give exhaustive reviews of the topic, otherwise you risk exhausting your reader! A well written introduction leaves readers satisfied that they understand why the experiment was done and what the author hoped to accomplish.

When you provide background information, be sure to **cite your sources**, especially when the information is not common knowledge (to determine what is common knowledge, see p. 44 in Chapter 3). Citing sources not only makes your statements more credible, it allows readers to find additional information on the topic of your paper.

Finally, write the opening sentence. This may be the most difficult part of the introduction to write. To help you do so, focus on your readers. Who are they and what are they likely to know about this topic? Come up with several opening sentences and then evaluate each sentence critically in terms of the level of your audience. Choose an opening sentence that is neither too simplistic nor too technical.

Tense In the course of providing background information on your topic, you will discuss scientific fact that is based on findings published in research papers. When describing scientific fact, use *present* tense.

> FACT: Peroxidase *is* completely denatured at temperatures above 80 °C (Duarte-Vázquez *et al.* 2003).

On the other hand, when stating the objectives of your study, use *past* tense. Past tense is preferred because proposing objectives is a completed action that you carried out before starting your actual study.

> OBJECTIVES: The purpose of this experiment *was* to determine the effect of temperature on peroxidase activity.

If the experiment lends itself to hypothesis testing, then state your hypothesis using a mixture of tenses. Notice how tenses are used in the following examples.

> HYPOTHESIS: We hypothesized [past tense] that enzyme activity will increase [future tense] with temperature up to a point.

HYPOTHESIS: We expected [past tense] enzyme activity to increase [present tense] with temperature up to a point.

HYPOTHESIS: Enzyme activity is or was [either present or past tense is appropriate] expected to increase with temperature up to a point.

Voice Active voice is preferred because it makes sentences shorter and more direct. But voice can also change the emphasis of a sentence, as illustrated by the following examples:

ACTIVE VOICE: Human activities are threatening the extinction of many species.

PASSIVE VOICE: The extinction of many species is threatened by human activities.

If your point is to emphasize *the role of human activities* in species extinctions, then active voice makes the stronger statement. If your focus is on *species extinction*, then passive voice may be more appropriate.

Effective Advertising

The whole point of writing your paper is to communicate your work to your fellow scientists. The abstract and the title are the primary tools potential readers will use to decide whether or not they are interested in your work.

Write the abstract

The abstract is a **summary of the entire paper** in 250 words or less. It contains:

- An introduction (scope and purpose)
- A short description of the methods
- The results
- Your conclusions

There are no literature citations or references to figures in the abstract.

After the title, the abstract is the most important part of the scientific paper used by readers to determine initial interest in the author's work. Abstracts are indexed in databases that catalogue the literature in the biological sciences. If an abstract suggests that the author's work may be relevant to your own work, you will probably want to read the whole article.

On the other hand, if an abstract is vague or essential information is missing, you will probably decide that the paper is not worth reading. When you write the abstract for your own laboratory report, put yourself in the position of the reader. If you want the reader to be interested in your work, write an effective abstract.

Writing the abstract is difficult because you have to condense your entire paper into 250 words or less. One strategy for doing this is to list the key points of each section, as though you were taking notes on your own paper. Then write the key points in full sentences. Revise the draft for clarity and conciseness using strategies such as using active voice, combining choppy sentences with connecting words, rewording run-on sentences, and eliminating redundancy. With each revision, look for ways to shorten the text so that the resulting abstract is a concise and accurate summary of your work.

The ability to write abstracts is important to a scientist's career. Should you someday wish to present your research at an academic society meeting, such as the Society for Neuroscience, the American Association for the Advancement of Science, or the National Association of Biology Teachers (to name just a few), you will be asked to submit an abstract of your presentation to the committee in charge of the meeting program. Your chances of being among the select field of presenters at these meetings are much better if you have learned to write a clear and intelligent abstract.

Write the title

The title is a **short, informative description** of the essence of the paper. You may choose a working title when you begin to write your paper, but revise the title after subsequent drafts. Remember that readers use the title to determine initial interest in the paper, so descriptive accuracy is the most essential element of your title. Brevity is nice if it can be achieved. Some journals (especially the British ones) are fond of puns and humor in their titles, but this kind of thing may be better left for later in your career.

Here are some examples of vague and undescriptive titles:

FAULTY: Quantitative Protein Analysis

FAULTY: The Assessment of Protein Content in an Unknown Sample

FAULTY: Egg White Protein Analysis

EXPLANATION: These titles leave the reader wondering what method of protein analysis was used and what sample was analyzed.

REVISION: Assessment of protein concentration in egg white using the biuret method

Here is another series of examples in which adding specific details improves the title:

FAULTY: Study of an Enzymatic Reaction

EXPLANATION: Specify the variables you studied. Specify the enzyme and the substrate in the reaction.

FAULTY: Initial velocity of enzymatic reactions under varying conditions

EXPLANATION: Was *more than one* enzymatic reaction studied? What were the *specific conditions*? If you only studied one reaction, use the singular.

REVISION: Effect of substrate and enzyme concentration and hydroxylamine (an inhibitor) on the initial velocity of the peroxidase–hydrogen peroxide reaction

Here is another example in which the title is made more descriptive by removing unnecessary words and adding the specific variable that was manipulated:

FAULTY: Explanation of seed germination in *H. vulgare*

EXPLANATION: Avoid using "filler phrases" such as "explanation of," "analysis of," and "study of." Give the common name and the scientific name of the organism for the reader's benefit. Focus on the specific aspect of seed germination that you studied.

REVISION: Effect of gibberellic acid concentration on starch remaining in the endosperm of barley (*Hordeum vulgare*) seeds

Documenting Sources

Whenever you use another person's ideas, whether they are published or not, you must document the source. This is done by citing the source in an abbreviated form in the text (**in-text reference**) and then giving the full reference in the References section at the end of the paper (**end reference**). An exception to this practice is personal communications, which are cited in the text, but are not listed among the end references. Only sources that have been cited in the text may be included in the References section.

The CSE Manual (2014) recommends using the Citation-Sequence System (C-S), the Citation-Name System (C-N), or the Name-Year System (N-Y) for documenting your sources. The system you actually use depends on your instructor's preference or on the format specified by the particular scientific journal in which you aspire to publish.

In all three systems, the *in-text reference* is intended to be inconspicuous. A superscripted number or a number in parentheses (C-S and C-N systems) or authors' names and year (N-Y system) are minimally disruptive to the flow of the sentence. Contrast this style with the lengthy introduction practiced in some disciplines in the humanities: "According to Warne and Hickock in their 1989 paper published in *Plant Physiology*, antheridiogen may be related to the gibberellins." **Do not use this style in your lab reports!**

Another difference between citations in the humanities and in scientific papers is that direct quotations are almost never used in the latter. Instead, write the information from the source text in your own words and then cite the source (see p. 45 in Chapter 3).

With regard to the *end reference*, the systems differ in the sequence of information and the listing of the month of publication. In the N-Y system, the year of publication follows the authors' names; in the C-S and C-N systems, the year follows the journal name. The month of publication is only used in the C-S and C-N systems.

The Name-Year system has the advantage that people working in the field will know the literature and, on seeing the authors' names, will understand the in-text reference without having to check the end reference. With the Citation-Sequence and Citation-Name systems, the reader must turn to the reference list at the end of the paper to gain the same information. Regardless of which system you use, learn the proper way to format both the in-text reference and the end reference and use one system consistently throughout any given paper.

Finally, do not list sources in the end reference list that you personally have not seen. If you feel that the original source is important enough to be cited, use the following approach:

Author (year) as cited by Author (year)

The Name-Year system

The *in-text reference* consists of author(s) and year. The author(s) may be cited in parentheses at the end of the sentence or they may be the subject of the sentence, as shown in the following examples:

TABLE 4.4 The number of authors determines how the source is cited in N-Y system

Number of Authors	Author as Subject	Parenthetical Reference (The comma between author[s] and year is optional.)
1	Author's last name (year) found that...	(Author's last name, year)
2	First author's last name and second author's last name (year) found that ...	(First author's last name and second author's last name, year)
3 or more	First author's last name followed by "and others" or *et al.* (year) found that ...	(First author's last name and others, year) or *et al.* instead of *and others*

Note: If you cite more than one paper published by the same author in *different* years, list them in chronological order: (Dawson 2001, 2003). If you cite more than one paper published by the same author in the *same* year, add a letter after the year: "...was described in recent work by Dawson (1999a, 1999b)."

PARENTHESES: C-fern gametophytes respond to antheridiogen only for a short time after inoculation (Banks and others 1993).

AS THE SUBJECT: Banks and others (1993) found that C-fern gametophytes respond to antheridiogen only for a short time after inoculation.

The **number of authors** determines how the *in-text reference* is written in the N-Y system (Table 4.4). For *one* author, write the author's last name and year. For *two* authors, write both authors' last names separated by the word *and* followed by the year. For *three or more* authors, write the first author's last name, the words *and others* (or *et al.*), and then the year.

In the *end references*, the sources are listed in **alphabetical order** according to the first author's last name. The format of the source determines which elements are included (Table 4.5). When there are 10 or fewer authors, list all authors' names. When there are more than 10 authors, list the first 10 and then write *et al.* or *and others* after the tenth name. For each reference, list the authors' names in the order they appear on the title page. Write each author's name in the form Last name First initials. Use a comma to separate one author's name from the next. Use a period only after the last author's name.

Examples of in-text references and their corresponding end references are given in Table 4.6. See The CSE Manual (2014) and Patrias (2007) for examples of many other kinds of sources.

TABLE 4.5 General format of two systems of source documentation used in scientific papers

Name-Year End Reference System	
The references are listed in **alphabetical order**. The last name is written first, followed by the initials. When there are 10 or fewer authors, list all authors' names. When there are more than 10 authors, list the first 10 and then write *et al.* or *and others* after the tenth name. Type references with hanging indent format.	
Journal article	First author's last name First initials, Subsequent authors' names separated by commas. Year of publication. Article title. Journal title Volume number(issue number): inclusive pages.
Article in book	First author's last name First initials, Subsequent authors' names separated by commas. Year of publication. Article title. In: Editors' names followed by a comma and the word *editors*. Book title, edition. Place of Publication: Publisher. pp inclusive pages.
Book	First author's or editor's last name First initials, Subsequent authors' or editors' names separated by commas. Year of publication. Title of book. Place of Publication: Publisher. Total number of pages in book followed by *p*.
Citation-Sequence End Reference System	
The references are listed **in the order they are cited**. The author's last name is written first, followed by the initials. When there are 10 or fewer authors, list all authors' names. When there are more than 10 authors, list the first 10 and then write *et al.* or *and others* after the tenth name.	
Journal article	Number of the citation. First author's last name First initials, Subsequent authors' names separated by commas. Article title. Journal title Year Month; Volume number(issue number): inclusive pages.
Article in book	Number of the citation. First author's last name First initials, Subsequent authors' names separated by commas. Article title. In: Editors' names followed by a comma and the word *editors*. Book title, edition. Place of Publication: Publisher; Year of publication. pp inclusive pages.
Book	Number of the citation. First author's or editor's last name First initials, Subsequent authors' or editors' names separated by commas. Title of book. Place of Publication: Publisher; Year of publication. Total number of pages in book followed by *p*.

The Citation-Sequence system

The *in-text reference* consists of a superscripted endnote (never a footnote) or a number in parentheses or square brackets within or at the end of the paraphrased sentence. The first reference cited is number 1, the second reference cited is number 2, and so on.

TABLE 4.6 Examples of in-text citation and end reference format of two systems of source documentation used in scientific papers

	Name-Year System
	IN-TEXT REFERENCES
3 or more authors 2 authors 1 author	Gametophytes of the tropical fern *Ceratopteris richardii* (C-fern) develop either as males or hermaphrodites. Their fate is determined by the pheromone antheridiogen (Näf 1979; Näf and others 1975). Banks and others (1993) found that gametophytes respond to antheridiogen only for a short time between 3 and 4 days after inoculation. Although the structure of antheridiogen is unknown, it is thought to be related to the gibberellins (Warne and Hickok 1989). Gibberellins are a group of plant hormones involved in stem elongation, seed germination, flowering, and fruit development (Treshow 1970).
	CORRESPONDING END REFERENCES
Journal article	Banks J, Webb M, Hickok L. 1993. Programming of sexual phenotype in the homosporous fern *Ceratopteris richardii*. Inter. J. Plant Sci. 154(4): 522-534.
Article in book	Näf U. 1979. Antheridiogens and antheridial development. In: Dyer AF, editor. The Experimental Biology of Ferns. London: Academic Press. pp. 436-470.
Journal article	Näf U, Nakanishi K, Endo M. 1975. On the physiology and chemistry of fern antheridiogens. Bot. Rev. 41(3): 315-359.
Book	Treshow M. 1970. Environment and Plant Response. New York: McGraw-Hill. 250 p.
Journal article	Warne T, Hickok L. 1989. Evidence for a gibberellin biosynthetic origin of *Ceratopteris* antheridiogen. Plant Physiol. 89(2): 535-538.
	Citation-Sequence System
	IN-TEXT REFERENCES
Sources are listed in the order they are cited	Gametophytes of the tropical fern *Ceratopteris richardii* (C-fern) develop either as males or hermaphrodites. Their fate is determined by the pheromone antheridiogen (1, 2). Gametophytes respond to antheridiogen only for a short time between 3 and 4 days after inoculation (3). Although the structure of antheridiogen is unknown, it is thought to be related to the gibberellins (4). Gibberellins are a group of plant hormones involved in stem elongation, seed germination, flowering, and fruit development (5).
	CORRESPONDING END REFERENCES
Article in book	1. Näf U. Antheridiogens and antheridial development. In: Dyer AF, editor. The Experimental Biology of Ferns. London: Academic Press; 1979. pp. 436-470.
Journal article	2. Näf U, Nakanishi K, Endo M. On the physiology and chemistry of fern antheridiogens. Bot. Rev. 1975 Jul-Sep; 41(3): 315-359.
Journal article	3. Banks J, Webb M, Hickok L. Programming of sexual phenotype in the homosporous fern *Ceratopteris richardii*. Inter. J. Plant Sci. 1993 Dec; 154(4): 522-534.
Journal article	4. Warne T, Hickok L. Evidence for a gibberellin biosynthetic origin of *Ceratopteris* antheridiogen. Plant Physiol. 1989 Feb; 89(2): 535-538.
Book	5. Treshow M. Environment and Plant Response. New York: McGraw-Hill; 1970. 250 p. **Pagination is optional. If present, this is the total number of pages in the book, not the pages from which information was cited.**

SUPERSCRIPTED ENDNOTE: There are four commonly used methods for determining protein concentration: the biuret method[1], the Lowry method[2], the Coomassie Blue (CB) dye-binding method[3], and the bicinchoninic acid (BCA) assay[4].

PARENTHESES: The Kjeldahl procedure is time-consuming and requires a large amount of sample (1, 2).

BRACKETS: Several review articles compare the advantages and disadvantages of these protein assays [5–10].

In the *end references*, the sources are listed in **numerical order** (in the order of citation). The format of the source determines which elements are included (Table 4.5). When there are 10 or fewer authors, list all authors' names. When there are more than 10 authors, list the first 10 and then write *et al.* or *and others* after the tenth name. Write each author's name in the form Last name First initials. Use a comma to separate one author's name from the next. Use a period only after the last author's name.

Examples of in-text references and their corresponding end references are given in Table 4.6. See The CSE Manual (2014) and Patrias (2007) for examples of many other kinds of sources.

The Citation-Name system

In the *end references*, the sources are listed in **alphabetical order** according to the first author's last name. The year and month of publication follow the journal name, as in C-S end reference format. The references are then numbered sequentially so that the first reference is number 1, the second reference is number 2, and so on. The *in-text references* consist of superscripted endnotes (never footnotes) or a number in parentheses or square brackets within or at the end of the paraphrased sentence.

Unpublished laboratory exercise

Unpublished material is usually not included in the References section. However, if your instructor asks that you cite laboratory exercises in your laboratory report, the *end reference* could look like this:

C-S: #. Author (omit if unknown). Title of lab exercise. Course number, Department, University. Year.

N-Y: Author (if unknown, replace with title of lab exercise). Year. Title of lab exercise. Course number, Department, University.

In N-Y format, the *in-text reference* for an unpublished lab exercise would include the author(s) and year, or, if the author is unknown, the title and year. The use of anonymous is not recommended (CSE Manual 2014).

Personal communication

Unpublished information obtained during a discussion or by attending a lecture should be acknowledged when you use it in your lab report or scientific paper. The in-text reference includes the authority, the date, and the words "personal communication" or "unreferenced." For example:

> Most viruses affecting honey bees have genomes composed of RNA rather than DNA (M. Pizzorno, personal communication, 2016 Sept 22).

It is **not necessary** to include personal communications in the references.

Internet sources

In the previous section you learned that the in-text reference and end reference format differs for journal articles, articles in a book, and books. These differences apply to both print and online publications. For a journal article, therefore, you should be able to locate on the website the names of the authors, a title, the journal name, a date of publication, the volume and issue number, and the extent (number of pages or similar). Besides this basic information, the CSE Manual (2014) recommends that you provide two additional items when your reference comes from the Internet: the URL (uniform resource locator) and the date accessed. **For your lab reports, it is sufficient to treat references obtained online as print sources** (unless your instructor tells you otherwise). If you would like to publish your work in a journal that adheres strictly to CSE guidelines, however, the following section shows the in-text reference and end reference format for an online journal article. For a comprehensive discussion of Internet citation formats along with many examples, see Patrias (2007).

The problem with sources on the Internet is that they may disappear at any time or their URL may change. To provide a persistent link to online articles and books, many publishers include a DOI (digital object identifier) on the first page of the publication (Figure 4.11). The DOI consists of a unique string of numbers and letters that, when pasted into a browser, leads directly to that publication. According to the APA (American Psychological Association) style manual, which is commonly used in the social sciences, when a DOI is given, the DOI rather than the URL should be included in the end reference. The goal is to get the reader to the source quickly and reliably.

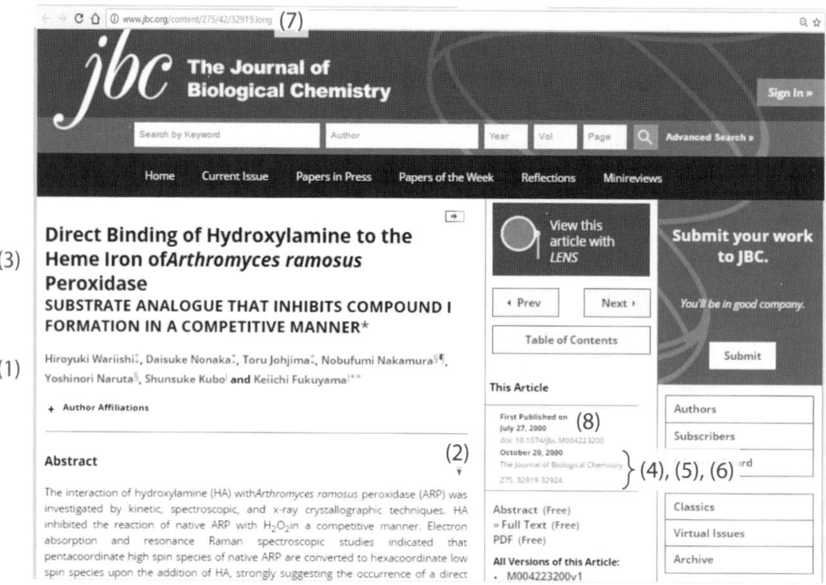

Figure 4.11 Web page for an online journal article. The basic information needed to cite a journal article includes (1) authors, (2) date of publication, (3) article title, (4) journal title, (5) volume and issue number (if given), and (6) inclusive pages. In addition, for an online journal article, include (7) the URL or (8) the DOI (digital object identifier) and the date accessed in the end reference.

> When URLs are used in text, they do *not* need special formatting. They do *not* need to be enclosed in angle brackets (< >) and they do *not* need to be underlined and in color. Every character in a URL is significant, as are spaces and capitalization. Very long URLs can be broken before a punctuation mark (tilde ~, hyphen -, underscore _, period ., forward slash /, backslash \, or pipe |). The punctuation mark is then moved to the next line.

Journal articles

The *in-text reference* for an online journal article is exactly the same as that for a printed journal article (see Table 4.6). A good approach for writing the *end reference* of an online journal article is to first locate the information you would need for a printed journal article, and then add the Internet-specific items (CSE Manual 2014). Choose one of the three systems—**Name-Year, Citation-Sequence**, or **Citation-Name**—and position the elements accordingly.

The general format for a *printed* end reference in the **Name-Year system**, including punctuation, is:

> Author(s). Date of publication. Title of article. Title of journal plus volume(issue): Inclusive page numbers

The corresponding format for an *online* reference, with the Internet information shown here in bold, is:

> Author(s). Date of publication. Title of article. Title of journal. **[date updated; date accessed]**; Volume(issue): Inclusive page numbers. **URL**

A screen shot of an online journal article web page is shown in Figure 4.11 and the elements required for citation are labeled. The corresponding end reference in **Name-Year format** with Internet-specific information is as follows:

> Wariishi H, Nonaka D, Johjima T, Nakamura N, Naruta Y, Kubo S, Fukuyama K. 2000. Direct binding of hydroxylamine to the heme iron of *Arthromyces ramosus* peroxidase. Substrate analogue that inhibits compound I formation in a competitive manner. *J Biol Chem*. [accessed 2016 Sept 22]; 275: 32919–32924. http://www.jbc.org/content/275/42/32919.long

The general format for a *printed* reference in the **Citation-Sequence** system, including punctuation, is:

> Number of the citation. Author(s). Title of article. Title of journal plus year and month; Volume(issue): Inclusive page numbers

The corresponding format for an *online* reference, with the Internet information shown here in bold, is:

> Number of the citation. Author(s). Title of article. Title of journal. Year and month **[date updated; date accessed];** Volume(issue): Inclusive page numbers. **URL**

The end reference for the same online journal article shown in Figure 4.11 in **Citation-Sequence format** would be:

> 1. Wariishi H, Nonaka D, Johjima T, Nakamura N, Naruta Y, Kubo S, Fukuyama K. Direct binding of hydroxylamine to the heme iron of *Arthromyces ramosus* peroxidase. Substrate analogue that inhibits compound I formation in a competitive manner. *J Biol Chem*. 2000 Oct [accessed 2012 Oct 29]; 275(42): 32919–32924. http://www.jbc.org/content/275/42/32919.long

Databases

A **database** is a collection of records with a standard format. Databases may be text-oriented or numerical and their content is usually accessed by means of a search box. You may cite an entire database if your goal is to make the reader aware of its existence, or you may only cite a part of the database to document an individual record. Some databases are available on paper and CD-ROM as well as on the Internet. Specify the medium, as Internet databases may contain more recent information than the corresponding paper or CD-ROM versions.

The general format for citing a database in the **Name-Year system** is:

> Title of Database [medium designator]. Beginning date – ending date (if given). Edition. Place of Publication: Publisher. [date updated; date accessed]. URL

To cite a database in the **Citation-Sequence system**, move the date after the publisher:

> Number of the citation. Title of Database [Medium Designator]. Edition. Place of Publication: Publisher. Beginning date – ending date (if given). [date updated; date accessed]. URL

As an example, the homepage of the BLAST database is shown in Figure 4.12. The nucleotide blast, protein blast, blastx, and tblastn databases are individual websites within the larger BLAST website. When citing websites within websites, the following rule applies: Always cite the most

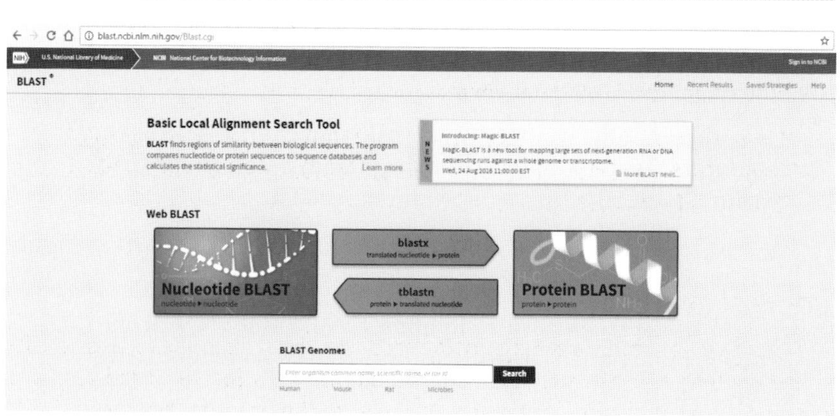

Figure 4.12 Homepage for the National Center for Biotechnology Information's BLAST database. To search for a specific nucleotide sequence, use nucleotide blast, one of the databases within the BLAST database.

specific organizational entity that you can identify (Patrias 2007). Database titles do not always follow the rules of English grammar and punctuation. Because they are proper nouns, however, reproduce the title as closely as possible to the format on the screen (maintain upper or lower case letters, run-together words, etc.). Sometimes the information needed for the reference may be absent or hard to find. In this example, the beginning to ending dates and the edition of the database are not specified. The location of the place of publication and the publisher are not given on this page, but can be found by clicking the **Contact** link at the bottom of the page. Do your best to reference the source with the information provided.

A good faith attempt at citing the nucleotide blast database in **Name-Year format** would be as follows:

> nucleotide blast [database on the Internet]. Bethesda (MD): U.S. National Library of Medicine, National Center for Biotechnology Information. [accessed 2016 Sept 20]. http://blast.ncbi.nlm.nih.gov/Blast.cgi

In Citation-Sequence format:

> 1. nucleotide blast [database on the Internet]. Bethesda (MD): U.S. National Library of Medicine, National Center for Biotechnology Information. [accessed 2016 Sept 20]. http://blast.ncbi.nlm.nih.gov/Blast.cgi

The *in-text reference* for a database in **Name-Year format** follows the same principles used for print publications (see Table 4.6) with a minor modification. The author is replaced with the title of the database and, when the date of publication is not known (as in the current example), the order of preference is the copyright date; the date of modification, update, or revision; and the date accessed (CSE Manual 2014). An example of an *in-text reference* in **Name-Year format** for this database would be:

> There was a 100% match between the DNA sequence of Sample 1 and the SV40 sequence in the NCBI databank (nucleotide blast database [accessed 2016]).

Homepages

A **homepage** is the main page of a website, which provides links to different content areas of the site. Most of the information required to cite a website is found on the homepage. Make sure the organization or individual responsible for the website is reputable and, if possible, confirm information on the site using another source. The general format for citing a homepage in **Name-Year format** is:

Title of Homepage. Date of publication. Edition. Place of publication: publisher; [date updated; date accessed]. URL

To cite a homepage in **Citation-Sequence format**, move the date after the publisher:

Number of the citation. Title of Homepage. Edition. Place of publication: publisher; date of publication [date updated; date accessed]. URL

An example of a homepage is shown in Figure 4.13. All of the information required to cite this source is readily located. When the date of publication is not specified, the order of preference is the copyright date; the

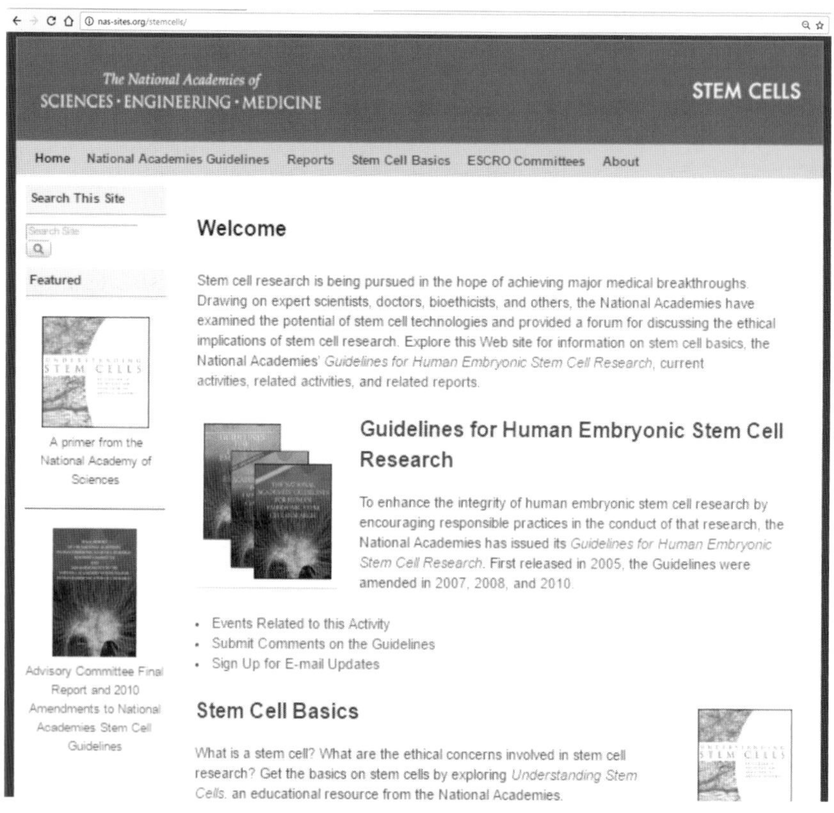

Figure 4.13 The National Academies' website for information on stem cells. Well-constructed homepages make it easy to find the title, date, responsible organization, and place of publication.

date of modification, update, or revision; and the date accessed (CSE Manual 2014). In this example, the copyright date, preceded by a lower case *c* is used in the end reference. The end reference in **Name-Year format**:

> Stem Cells at the National Academies. c2016. Washington DC: National Academy of Sciences; [accessed 2016 Sept 20]. http://nas-sites.org/stemcells/

The end reference in Citation-Sequence format:

> 1. Stem Cells at the National Academies. Washington DC: National Academy of Sciences; c2016 [accessed 2016 Sept 20]. http://nas-sites.org/stemcells/

The *in-text reference* for a homepage in **Name-Year format** follows the same principles used for print publications (see Table 4.6) with a minor modification. The author is replaced with the title of the homepage. For the year, the order of preference is the date of publication; the copyright date; the date of modification, update, or revision; and the date accessed (CSE Manual 2014).

Emails and discussion lists

Electronic mail (email) and discussion lists (LISTSERVs, news groups, bulletin boards, etc.) are usually considered to be a form of personal communication (see pp. 94–95). Information obtained through personal communication is cited in the in-text reference, but not in the end reference list. The in-text reference should include the name of the authority, the date, and the words "personal communication" or "unreferenced" to indicate that the citation is not listed in the References section.

Go to the **COMPANION WEBSITE** • **sites.sinauer.com/knisely5e**
for samples, template files, and tutorial videos

Chapter 5

Revision

Revision—reading your paper and making corrections and improvements—is an important task that usually does not get nearly the attention it deserves. Too many students write the first draft of their laboratory report the night before it is due and hand in the hard copy, still warm from the printer, without even having proofread it.

The truth is, most writers cannot produce a clear, concise, and error-free product on the first try. It may take several revisions before the writer is satisfied that he or she has conveyed, with clarity and logic, the motivation for writing the paper, the important findings, and the conclusions. Do not try to write and revise your entire paper in one marathon session. Instead, **break up the writing process** into multiple, shorter segments. The breaks give your mind time to process what you've written. Starting to write early also allows you to get help if necessary, get feedback from your peer reviewer, and make final revisions.

Most excellent writers were not born that way. They achieved excellence through "deliberate practice" (Martin 2011). The old adage "practice makes perfect" applies not just to musicians and athletes, but to *you* as an aspiring author in the biological sciences. So if writing doesn't come naturally to you, take heart. Writing laboratory reports becomes easier with practice, especially if you learn from your mistakes.

Getting Ready to Revise

Take a break

The first step in revision is *not* to do it immediately after you have completed the first draft. You need to distance yourself from the paper to gain the objectivity needed to read the paper critically. So take a break, and go for a run or get a good night's sleep.

Slow down and concentrate

Find a quiet room where you won't be disturbed. Don't read your paper the same way you wrote it. Instead, **change its appearance** by increasing the zoom level or converting the Word document to a PDF (LR Communication Systems 1999; Corbett 2011). Even better, read a printed copy. Next, use some of your other senses to force yourself to **slow down**. For example, reading aloud involves the sense of hearing. Pointing to each word with your finger adds the sense of touch. If something doesn't sound right, trust your instincts. Figure out what is bothering you and fix the troublesome passage. Finally, **don't try to fix every kind of error in one pass**. If you do, you're sure to miss some.

Think of your audience

The rest of this chapter describes a systematic approach to revising your writing, whether for a lab report, a poster, or an oral presentation. Remember for whom you are writing and keep in mind the needs and motivations of your audience. Revise your writing with the goal of meeting or exceeding their expectations in a style that eases comprehension.

Editing

Revision can be divided into two stages: editing and proofreading. **Editing** is done first and involves reading for content and organization. The editing process proceeds from the broad to the specific. First, evaluate the overall structure of your paper. Then, read each individual section, paragraph, sentence, and word critically. And don't forget to check that the data were plotted correctly and that the description of each visual in the text is accurate. After editing is completed, **proofread** the paper. This involves correcting errors in spelling, punctuation, grammar, and overall format. This chapter provides specific guidance for each step of the revision process. The "**Laboratory Report Errors**" section on pp. 149–161 illustrates common errors to look out for when you revise your lab report.

Evaluate the overall structure

If your instructor provided a rubric for your lab report, use it as a checklist for content and organization. You may also wish to print out the "Biology Lab Report Checklist" on pp. 138–140 from <http://sites.sinauer.com/Knisely5E>. If you are preparing a paper for submission to a journal or conference, follow the relevant *Instructions to Authors*.

Most scientific papers are divided into **standard sections**: title, authors' name(s), abstract, *I*ntroduction, *M*aterials & Methods, *R*esults, *D*iscussion, and references (known as IMRD format). Readers of scientific papers like

TABLE 5.1 Checklist for section content

Section	Content
Title	Contains keywords that describe the essence of the study.
	"Filler" phrases like "Analysis of the...," or "Study of the..." are not used.
Abstract	Contains an introduction, a brief description of the methods, results, and conclusions.
Introduction	The structure is broad to specific: The main topic is introduced on a general level. The question or unresolved problem is stated. The objectives of the current study are presented in the last paragraph.
	Background information is provided by citing published sources.
Materials and Methods	Contains all of the relevant details to enable another trained scientist to repeat the procedure.
	Routine procedures are not described.
Results	Contains text and visuals. The descriptive text precedes each visual and includes a reference to the figure or table number.
	The text describes the results shown in each visual. The results are not explained or interpreted.
	Visuals include graphs, tables, photos, gel images, and so on, which contain the numerical or descriptive data. Each visual has its own caption that begins with Figure (or Table) and a number followed by a title that can be understood apart from the text.
Discussion	The structure is specific to broad: The results are explained and interpreted. The results are compared to those in other studies, usually published journal articles. If warranted, there may be a discussion of why the results in the current study are important or how these results contribute to our understanding of the topic.
References	The full reference is given for each source cited in the text.
	References that have not been cited are not listed.

this format because they know where to look to find certain kinds of information. Confirm that these sections are in the **right order** in your paper. Then check that each section has the **appropriate content** (Table 5.1) by underlining each component on the printed pages or highlighting them on the computer screen (Hofmann 2014).

Do the math at least twice

Double-check your calculations and spreadsheet data entries. A mistake at this stage will have a negative domino effect, resulting in inaccurate figures or tables and a faulty discussion and interpretation of the results.

Organize each section

The phrases you underlined when you checked for content should provide a rough outline of each section. Do the **most important topics stand out**? Does the **order of the topics make sense** chronologically or sequentially? Is the order what your audience expects (for example, are the topics arranged from broad to specific in the Introduction section and specific to broad in the Discussion section)? Rearrange paragraphs so that the important topics can be identified easily, in an order that makes sense.

Make coherent paragraphs

Each paragraph should focus on only one topic Make the topic sentence the first sentence in the paragraph. Follow the topic sentence with supporting sentences that directly relate to the main idea.

Arrange the sentences in a logical order Different strategies can be used depending on the section of the scientific paper. For example, in the Materials and Methods section, it makes sense to describe the procedure *sequentially* (in the order the steps were carried out).

> EXAMPLE: Barley seeds were surface sterilized with 10% bleach. Then they were cut in half, keeping the endosperm portion and discarding the embryo portion. Each half was placed cut side down on three pieces of sterile filter paper that had been soaked in 3.5 mL of HEPES-EGTA-Ca^{2+} buffer or a certain concentration of hormone solution.

Chronological sentence order might be used in the Introduction section to describe the sequence of events leading up to our current state of knowledge about a topic.

> EXAMPLE: Germination begins when a seed imbibes water. Early studies on barley seeds showed that the hormone gibberellic acid (GA) is involved in this process (Paleg 1960; Yomo 1960). Varner (1964) demonstrated that the enzymes that digest starch into sugars are produced in the aleurone layer. For the past 5 decades, it was suspected that GA binds to a receptor on the surface of the aleurone cells, but how the protein subunits in the receptor function in signal transduction was discovered only recently (Ueguchi-Tanaka et al. 2005).

General-to-specific is another way to arrange sentences in the introduction. In this approach, the paragraph starts with a general idea that is then supported by details or examples.

EXAMPLE: Germination begins when a seed imbibes water. Water triggers the release of gibberellic acid (GA) from the embryo. The hormone diffuses through the endosperm and binds to receptors on the plasma membrane of the aleurone cells. Through a signal transduction pathway that is not fully understood, digestive enzymes are activated and released into the endosperm. One of those enzymes is α-amylase, which metabolizes starch into its sugar subunits.

On the other hand, *specific-to-general* is the order expected in the Discussion section. As shown in the following example, the paragraph begins with a brief recapitulation of a specific result. Subsequent sentences explain the result based on our current understanding of the topic.

EXAMPLE: Embryoless half-seeds exposed to higher concentrations of gibberellic acid had a lower percentage of starch remaining after one week (Figure 2). Gibberellic acid is known to induce the production of α-amylase, an enzyme that hydrolyzes starch (Lovegrove and Hooley 2000). During germination, gibberellic acid, produced by the embryo, binds to receptors in the aleurone layer and activates α-amylase, which then degrades the starch in the endosperm into glucose, which provides the energy for the embryo to grow.

Use signal words and phrases to guide readers from one sentence to the next Signal words and phrases (also called **transitions**) help readers see relationships between sentences. Smooth transitions help readers to follow the writer's thought process, thereby increasing comprehension.

Consider the following example:

FAULTY: Catalase is an enzyme that breaks down hydrogen peroxide in both plant and animal cells. Low or high temperature can lower the rate at which the catalase can react with the hydrogen peroxide. In optimal conditions, the enzyme functions at a rate that will prevent any substantial buildup of the toxin. If the temperature is too low, the rate will be too slow, but high temperatures lead to the denaturation of the enzyme.

Where is the writer going with this paragraph? The sentences do not seem to flow because there is no guidance from the writer on how one sentence is related to the next. To improve flow, use signal words and phrases such as *however, thus, although, in contrast, similarly, on the other hand, in addition to,* and *furthermore.* Signal words and phrases may also be key words that are repeated from one sentence to the next. Notice how the addition of transitions guides the reader step by step through this passage.

REVISION: Catalase is an enzyme that breaks down hydrogen peroxide in both plant and animal cells. One of the factors that affects the rate *of this reaction* is temperature. At optimal *temperatures,* the rate is sufficient to prevent substantial buildup of the toxic *hydrogen peroxide.* If the temperature is too low, *however,* the rate will be too slow, and hydrogen peroxide *accumulates* in the cell. *On the other hand,* high temperatures may denature the *enzyme.*

Write meaningful sentences

Each sentence should **say something meaningful** and not repeat what was said before (**avoid redundancy**). Consider the following examples.

EXAMPLE: ~~From the data that has been gathered, a graph depicting the effect of various pH environments on the rate of catalase activity is represented. The graph displays the pH tested and the reaction rate. The data plotted is an accumulation of data from several lab sections. Through analyzing the graph I can see that~~ Tthere ~~is~~ was no activity below a pH of 4 or above 10. Maximum catalase activity occurred at pH 7.

EXPLANATION: The first three sentences do not convey anything substantive about the *results.* Only the last two sentences contain meaningful information.

EXAMPLE: As the enzyme concentration increased, the initial velocity increased ~~as well~~ (Figure 3). ~~There is an overall gradual increasing relationship between the enzyme concentration and the initial velocity. This increasing relationship seems to remain constant from the lowest concentration of 0 mM to the highest concentration which is 2 mM. This graph showed that there was a fairly increasing positive linear relationship between enzyme concentration and initial velocity.~~

EXPLANATION: The last three sentences say the same thing as the first sentence. The only substantive piece of information missing from the first sentence is that the trend was linear. Eliminate repetition and describe the results in as few words as possible.

To edit redundant sentences, take the best parts of those sentences and combine them into one concise sentence. Put yourself in your reader's shoes: Would you rather waste precious minutes wading through verbiage or get needed information with minimal effort?

Technical accuracy Sentences that provide background information on a topic (as in the Introduction section), describe procedures (in the Materials and Methods section), or explain results (in the Discussion section) should be based on scientific fact. When in doubt, check your references, including secondary sources such as your textbook. Furthermore, make sure that your description of the results shown in each visual is accurate. In particular, pay attention to words like *increase* and *decrease*. Check that you did not mix up the results when you describe multiple data sets plotted on one graph.

Sentence length Short sentences that contain only one idea are easy to understand. A text that contains nothing but short sentences, however, may be perceived as childish at best or hard to follow at worst. On the other hand, long, needlessly complex sentences obscure the main idea and slow comprehension. Aim for a mixture of short and long sentences in your writing. Use more words to explain a complex idea, but keep each sentence focused on just one idea.

Here are some examples of **needlessly complex sentences**.

FAULTY 1: *There are* two protein assays *that* are often used in research laboratories.

REVISION: Two protein assays are often used in research laboratories. (Avoid unnecessary words and phrases that "pad" a sentence.)

FAULTY 2: *It is interesting to note that* some enzymes are stable at temperatures above 60 °C.

REVISION: Some enzymes are stable at temperatures above 60 °C. (Avoid unnecessary introductory phrases.)

FAULTY 3: *The analyses were done on* the recombinant DNA to determine which piece of foreign DNA was inserted into the vector.

REVISION: The recombinant DNA was analyzed to determine which piece of foreign DNA was inserted into the vector. (Make *DNA*, not the *analyses*, the subject of the sentence.)

FAULTY 4: *We make the recommendation* that micropipettors be used to measure volumes less than 1 mL.

REVISION: We recommend that micropipettors be used to measure volumes less than 1 mL. (Replace sluggish noun phrases [nominalizations] with verb phrases.)

FAULTY 5: These assays alone cannot *tell* what the protein concentration of a substance is.

REVISION: These assays alone cannot determine the protein concentration of a substance. (Replace colloquial expressions with precise alternatives.)

Emphasize the subject Putting the subject at the beginning of the sentence makes the subject stand out. Position the verb close by so that there is no doubt about the subject's action.

Active and passive voice In **active voice**, the subject *performs* the action. In **passive voice**, the subject *receives* the action. Consider the following example:

PASSIVE: The clam was opened by the sea star. (Emphasis on *clam*)

ACTIVE: The sea star opened the clam. (Emphasis on *sea star*)

Although the meaning is the same in both sentences, notice the difference in emphasis. In active voice, the emphasis is on the performer, and the action takes place in the direction the reader reads the sentence. Active voice is recommended by most style guides for reasons that include the following:

- It sounds more natural and is easier for the reader to process.
- It is shorter and more dynamic.
- There is no ambiguity about who/what the subject of the sentence is, or about who did the action.

Consider the following example:

PASSIVE: It was concluded from this observation that...

ACTIVE: I concluded from this observation that...

Passive voice leaves the reader wondering who drew the conclusion; active voice conveys this information clearly.

While active voice is generally preferred, passive voice may be more appropriate when *what* is being done is more important than *who* is doing it. For example:

PASSIVE: Catalase was extracted from a turnip. (Emphasis on *catalase*)

ACTIVE: I extracted catalase from a turnip. (Emphasis on *I*)

Notice the difference in emphasis. Is it really important to the success of the procedure that *you* did it, or does the emphasis belong on the catalase?

A paper that contains a mixture of active and passive voice is pleasant to read. Your decision to use active or passive voice in a sentence should ultimately be determined by clarity and brevity. In other words, use active voice to emphasize the subject and the fact that the subject is performing the action. Use passive voice when the action is more important than who is doing it.

Present or past tense In scientific papers, present tense is used mainly in the following situations:

- To make generally accepted statements (for example, "Photosynthesis *is* the process whereby green plants produce sugars").
- When referring directly to a table or figure in your paper (for example, "Figure 1 *is* a schematic diagram of the apparatus").

- When stating the findings of published authors (for example, "Catalase HPII from *E. coli is* highly resistant to denaturation [Switala and others 1999]").

Past tense is used mainly in the following situations:

- To report your own work, especially in the abstract, Materials and Methods, and Results sections, because it remains to be seen if the scientific community accepts your work as fact (for example, "At temperatures above 37 °C, catalase activity *decreased* (Figure 3)").
- To cite another author's findings directly (for example, "Miller and others (1998) *found* that…").

Choose your words carefully

Words are the basic organizational unit of language. The words you choose and how you arrange them in a sentence will determine how well you convey your message to your readers. Beware of the following word-level problems:

Keep related words together Consider the following sentence taken from an English-language newspaper in Japan: "A committee was formed to examine brain death in the Prime Minister's office." Although brain death in the Prime Minister's office may be a political reality, what was really intended was, "A committee was formed in the Prime Minister's office to examine brain death."

Redundancy **Redundancy** means using two or more words that mean the same thing. This problem is easily corrected by eliminating one of the redundant words (Table 5.2). Along with empty phrases, redundancy is a source of **wordiness**, using too many words to convey an idea.

Some people think that using more words makes them sound important. In science, however, wordiness should be avoided at all costs, because it indicates that the writer can't communicate clearly. For student writers whose papers are evaluated by instructors, lack of clarity translates into a

TABLE 5.2 Examples of redundancy

Redundant	Revised
It is absolutely essential…	It is essential…
mutual cooperation	cooperation
basic fundamental concept	basic concept or fundamental concept
totally unique	unique
The solution was obtained and transferred…	The solution was transferred…

low grade. Researchers and faculty members, whose reputation depends on the number and quality of their publications, simply cannot afford *not* to write clearly, because poorly written papers may be equated with shoddy scientific methods.

Empty phrases Replace empty phrases with a concise alternative (Table 5.3). Put yourself in your reader's shoes. Which of the following two sentences (inspired by VanAlstyne 2005) would you rather read?

FAULTY: It is absolutely essential that you use a minimum number of words in view of the fact that your reader has numerous other tasks to complete at the present time.

REVISION: Write concisely, because your reader is busy.

Initially it is difficult to write in (and read) the terse, get-to-the-point style that characterizes scientific papers. With practice, however, you may come to appreciate this style because in a well-written paper, not a word is wasted. The benefit to you as a reader is that you extract a maximum amount of information from a minimum amount of text.

Ambiguous use of *this*, *that*, and *which* Ambiguity results when *this*, *that*, or *which* could refer to more than one subject.

FAULTY: The data show that the longer the enzyme was exposed to the salt solution, the lower the enzyme activity in the assay. *This* means that the salt changes the conformation of the enzyme, *which* makes it less reactive with the substrate.

EXPLANATION: The subject of *this* and *which* is unclear.

REVISION: The longer the enzyme was exposed to the salt solution, the lower the enzyme activity in the assay. Exposure to the salt solution may change the conformation of the enzyme, resulting in lower enzyme-substrate activity.

Ambiguous use of pronouns (him, her, it, he, she, its) Ambiguity results when a pronoun could refer to two possible antecedents.

TABLE 5.3 Examples of empty phrases

Empty	Concise
a downward trend	a decrease
a great deal of	much higher
a majority of	most
accounted for the fact that	because
as a result	so, therefore
as a result of	because
as soon as	when
at which time	when
at all times	always
at a much greater rate than	faster
at the present time, at this time	now, currently
based on the fact that	because
brief in duration	short, quick
by means of	by
came to the conclusion	concluded
despite the fact that, in spite of the fact that	although, though
due to the fact that, in view of the fact that	because
for this reason	so
in fact	*omit this phrase*

FAULTY: With time, salt changes the conformation of the enzyme, which makes *it* less reactive with the substrate.

EXPLANATION: *It* could refer to *salt* or *enzyme*. To eliminate the ambiguity, replace *it* with the appropriate noun phrase.

REVISION: With time, salt changes the conformation of the enzyme, so that the enzyme can no longer react with its substrate.

Word usage When you use the right words in the right situations, readers have confidence in your work. Use a standard dictionary whenever you are not sure about word usage. Consult your textbook and

TABLE 5.3 *(continued)*

Empty	Concise
functions to, serves to	*omit this phrase*
degree of	higher, more
in a manner similar to	like
in the amount of	of
in the vicinity of	near, around
is dependent upon	depends on
is situated in	is in
it is interesting to note that, it is worth pointing out that	*omit these kinds of unnecessary introductions*
it is recommended	I (we) recommend
on account of	because, due to
prior to	before
provided that	if
referred to as	called
so as to	to
through the use of	by, with
with regard to	on, about
with the exception of	except
with the result that	so that

laboratory exercise for proper spelling and usage of technical terms. The following word pairs are frequently confused in students' lab reports.

absorbance, absorbency, observance *Absorbance* is how much light a solution absorbs; absorbance is measured with a spectrophotometer. *According to Beer's law, absorbance is proportional to concentration. Absorbency* is how much moisture a diaper or paper towel can hold. *Brand A paper towels show greater absorbency than Brand B paper towels. Observance* is the act of observing. *Government offices are closed today in observance of Independence Day.*

affect, effect *Affect* is a verb that means "to influence." *Temperature affects enzyme activity. Affect* is rarely used as a noun in biology, although it has a specific meaning in psychology. *Effect* can be used either as a noun or a verb. When used as a noun, *effect* means "result." *We studied the effect of temperature on enzyme activity.* When used as a verb, effect means "to cause." *High temperature effected a change in enzyme conformation, which destroyed enzyme activity.*

alga, algae See plurals.

amount, number Use *amount* when the quantity cannot be counted. *The reaction rate depends on the amount of enzyme in the solution.* Use *number* if you can count individual pieces. *The reaction rate depends on the number of enzyme molecules in the solution.*

analysis, analyses See plurals.

bacterium, bacteria See plurals.

bind, bond *Bind* is a verb meaning "to link." *The active site is the region of an enzyme where a substrate binds. Bond* is a noun that refers to the chemical linkage between atoms. *Proteins consist of amino acids joined by peptide bonds. Bond* used as a verb means "to stick together." *This 5-minute epoxy glue can be used to bond hard plastic.*

complementary, complimentary *Complementary* means "something needed to complete" or "matching." *The DNA double helix consists of complementary base pairs: A always pairs with T, and G with C. Complimentary* means "given free as a courtesy." *The brochures at the visitor's center are complimentary.*

confirmation, conformation *Confirmation* means "verification." *I received confirmation from the postal service that my package had arrived. Conformation* is the three-dimensional structure of a macromolecule. *Noncovalent bonds help maintain a protein's stable conformation.*

continual, continuous *Continual* means "going on repeatedly and frequently over a period of time." *The continual chatter of a group of inconsiderate students during the lecture annoyed me. Continuous* means "going on without interruption over a period of time." *The bacteria were grown in L-broth continuously for 48 hr.*

create, prepare, produce *Create* is to cause to come into existence. *The artist used wood and plastic to create this sculpture. Prepare* means "to make ready." *The protein standards were prepared from a 50 mg/mL stock solution. Produce* means to make or manufacture. *The reaction between hydrogen peroxide and catalase produces water and oxygen.*

datum, data See plurals.

different, differing *Different* is an adjective that means "not alike." An adjective modifies a noun. *Different concentrations of bovine serum albumin were prepared. Differing* is the intransitive tense of "to differ," a verb that means "to vary." It is incorrect to replace the word *different* with *differing* in the preceding example, because *differing* implies that a single concentration changes depending on time or circumstance. This situation is highly unlikely with bovine serum albumin, which is quite stable under laboratory conditions! An acceptable use of *differing* is shown in the following example: *Bovine serum albumin solutions, differing in their protein content, were prepared.*

effect, affect See affect, effect.

fewer, less Use *fewer* when the quantity can be counted. *The reaction rate was lower, because there were fewer collisions between enzyme and substrate molecules.* Use *less* when the quantity cannot be counted. *The weight of this sample was less than I expected.*

formula, formulas, formulae See plurals.

hypothesis, hypotheses See plurals.

its, it's *Its* is a possessive pronoun meaning "belonging to it." *The Bradford assay is preferred because of its greater sensitivity. It's* is a contraction of "it is." *The Bradford assay is preferred because it's more sensitive.* (Note: Contractions should not be used in formal writing.)

less, fewer See fewer, less.

lose, loose *Lose* means to misplace or fail to maintain something. *An enzyme may lose its effectiveness at high temperatures. Loose* means "not tight." *When you autoclave solutions, make sure the lid on the bottle is loose.*

lowered, raised Both of these are transitive verbs, which means that they require a direct object (a noun to act on). **Wrong**: *The fish's body temperature lowered in response to the cold water.* **Right**: *The cold water lowered the fish's body temperature.*

media, medium See plurals.

observance See absorbance, absorbency, observance.

phenomenon, phenomena See plurals.

plurals The plural and singular forms of some words used in biology are given in Table 5.4. A common mistake with these words is not making the subject and verb agree. Some disciplines treat *data* as singular, but scientists subscribe to the strict interpretation that *data* is plural. The data *show*… (not *shows*) is correct.

prepare See create, prepare, produce.

produce See create, prepare, produce.

raised, lowered See lowered, raised.

ratio, ration *Ratio* is a proportion or quotient. *The ratio of protein in the final dilution was 1:5. Ration* is a fixed portion, often referring to food. *The Red Cross distributed rations to the refugees.*

TABLE 5.4 Singular and plural of words frequently encountered in biology

Singular	Plural
alga	algae
analysis	analyses
bacterium	bacteria
criterion	criteria
datum (rarely used)	data
formula	formulas, formulae
hypothesis	hypotheses
index	indexes, indices
medium	media
phenomenon	phenomena
ratio	ratios

strain, strand A *strain* is a line of individuals of a certain species, usually distinguished by some special characteristic. *The lacI⁻ strain of E. coli produces a nonfunctional repressor protein.* A *strand* is a ropelike length of something. *The strands of DNA are held together with hydrogen bonds.*

than, then *Than* is an expression used to compare two things. *Collisions between molecules occur more frequently at high temperatures than at low temperatures. Then* means "next in time." *First 1 mL of protein sample was added to the test tube. Then 4 mL of biuret reagent was added.*

that, which Use *that* with restrictive clauses. A restrictive clause limits the reference to a certain group. Use *which* with nonrestrictive clauses. A nonrestrictive clause does not limit the reference, but rather provides additional information. Commas are used to set off nonrestrictive clauses but not restrictive clauses. Consider the following examples:

EXAMPLE 1: The Bradford assay, which is one method for measuring protein concentration, requires only a small amount of sample. (*Which* begins a phrase that provides additional information, but is not essential to make a complete sentence.)

EXAMPLE 2: Enzyme activity decreased significantly, which suggests that the enzyme was denatured at 50 °C. (*Which* refers to the entire phrase *Enzyme activity decreased significantly,* not to any specific element.)

EXAMPLE 3: The samples that had high absorbance readings were diluted. (*That* refers specifically to *the samples.*)

various, varying *Various* is an adjective that means "different." *Various hypotheses were proposed to explain the observations. Varying* is a verb that means "changing." *Varying the substrate concentration while keeping the enzyme concentration constant allows you to determine the effect of substrate concentration on enzyme activity.* Analogous to *different, differing,* replacing the word *various* with *varying* in the preceding example changes the meaning of the sentence. *Varying* implies that a single hypothesis changes depending on time or circumstance. *Various* implies that different hypotheses were proposed.

Jargon and scientific terminology **Jargon** refers to words and abbreviations used by specialists. Whenever you use terms that may be unfamiliar to your audience, define them. Always write out the full expression when first using an abbreviation. Scientific words that you learned in class are *not* jargon and should be used in your writing. When you use scientific terminology correctly, your readers have confidence in your knowledge.

Clichés and slang **Clichés** are tired, worn out expressions that have no place in an exciting field like biology. Slang should not be used, because colloquial language is not appropriate in formal, written assignments like lab reports.

Gender-neutral language Years ago, it was customary, for the sake of simplicity, to use masculine pronouns to refer to antecedents that could be masculine or feminine, but that use of language is no longer accepted.

SEXIST: The clarity with which a biology student writes *his* lab reports affects *his* grade.

This practice is no longer considered to be politically correct. One solution that preserves equality, but makes sentences unnecessarily complex, is to include both masculine and feminine pronouns, as in the following example.

EQUAL BUT AWKWARD: The clarity with which a biology student writes *his or her* lab reports affects *his or her* grade.

Two better alternatives are to make the antecedent plural (revision 1) or to rewrite the sentence to avoid the gender issue altogether (revision 2).

REVISION 1: The clarity with which biology students write *their* lab reports affects *their* grade.

REVISION 2: *Writing clearly* has a positive effect on a biology student's grade. (Change the subject from *biology* student to *writing clearly*.)

Construct memorable visuals

Visuals often make the difference in how well you convey your message to your readers or listeners. Make sure you use the **appropriate visual** for the data (see pp. 60–75 in Chapter 4). Make sure **every visual serves a purpose**, because unnecessary visuals only dilute the significance of your message. Check that the **visuals are positioned in the right order** and that **each visual is described** in the text.

Proofreading: The Home Stretch

Proofreading is the last stage of revision. Like editing, it requires intense focus and slow, careful reading to find errors in format, spelling, punctuation, and grammar. **Grammar** refers to the rules that deal with the form and structure of words and their arrangement in sentences. See Hacker and Sommers (2016), Bullock et al. (2014), or Lunsford (2013, 2015, 2016) for a more comprehensive treatment of the subject.

Make subjects and verbs agree

We learn early on in our formal education to make the verb agree with the subject. Most of us know that *the sample was…*, but that *the samples were….* Most errors with subject–verb agreement occur when there are words *between* the subject and the verb, as in the following example.

EXAMPLE: The *kinetic energy* of molecules *is* (not *are*) lower at 6 °C than 45 °C.

EXAMPLE: The *chance* of collisions between enzyme and substrate molecules *increases* (not *increase*) under those conditions.

EXAMPLE: The enzyme has a *range* of temperatures that *is* (not *are*) optimal for activity.

When you write complex sentences, ask yourself what the subject of the sentence is. Look for the verb that goes with that subject. Then, mentally remove the words in between the two, and make the subject and its verb agree.

A second situation in which subject–verb agreement becomes confusing is when there are two subjects joined by *and*, as in the following example.

EXAMPLE: An enzyme's amino acid *sequence and* its three-dimensional *structure make* (not *makes*) the enzyme–substrate relationship unique.

Compound subjects joined by *and* are almost always plural.

A third situation involving subject–verb agreement is that when numbers are used in conjunction with units, the *quantity* is considered to be *singular*, not plural.

EXAMPLE: To extract the enzyme, 12 g of turnips *was* (not *were*) homogenized with 150 mL of cold, distilled water.

Write in complete sentences

A complete sentence consists of a subject and a verb. If the sentence starts with a subordinate word or words such as *after, although, because, before, but, if, so that, that, though, unless, until, when, where, who,* or *which,* however, another clause must complete the sentence.

FAULTY 1: High temperatures destroy the three-dimensional structure of enzymes. Thus changing the effectiveness of the enzymes. (The second "sentence" is a fragment.)

REVISION: High temperatures destroy the three-dimensional structure of enzymes, thus changing their effectiveness. (Combine the fragment with the previous sentence, changing punctuation as needed.)

FAULTY 2: The standard curve for the biuret assay was used to determine the protein concentration of the serial dilutions of the egg white. Although only those dilutions whose protein concentrations fell within the sensitivity range of the assay were multiplied by the dilution factor to give the original concentration of the egg white. (The second "sentence" is a fragment.)

REVISION: The standard curve for the biuret assay was used to determine the protein concentration of the serial dilutions of the egg white. Only those dilutions whose protein concentrations fell within the sensitivity range of the assay were multiplied by the dilution factor to give the original concentration of the egg white. (Delete the subordinate word[s] to make a complete sentence.)

Revise run-on sentences

Run-on sentences consist of two or more independent clauses joined without proper punctuation. Each independent clause could stand alone as a complete sentence. Run-on sentences are common in first drafts, where your main objective is to get your ideas down on paper (or electronic media, if you use a computer). When you revise your first draft, however, use one of the following strategies to revise run-on sentences:

- Insert a comma and a coordinating conjunction (*and, but, or, nor, for, so,* or *yet*).
- Use a semicolon or possibly a colon.
- Make two separate sentences.
- Rewrite the sentence.

FAULTY 1: The class data for the Bradford method were scattered, those for the biuret method were closer.

REVISION A: The class data for the Bradford method were scattered, but those for the biuret method were closer. (Use a coordinating conjunction.)

REVISION B: The class data for the Bradford method were scattered; those for the biuret method were closer. (Use a semicolon.)

FAULTY 2: The readings from the spectrophotometer should show a correlation between protein concentration and absorbance, this is Beer's law, which relates absorbance to the path length of light along with molar concentration of a solute and the molar coefficient. (Fused sentence.)

REVISION A: The readings from the spectrophotometer should show a correlation between protein concentration and absorbance; this is Beer's law, which relates absorbance to the path length of light along with molar concentration of a solute and the molar coefficient. (Use a semicolon to separate the two clauses.)

REVISION B: The readings from the spectrophotometer should show a correlation between protein concentration and absorbance. This relationship is described by Beer's law, which relates absorbance to the path length of light along with molar concentration of a solute and the molar coefficient. (Make two separate sentences.)

FAULTY 3: An increase in enzyme concentration increased the reaction rate as did an increase in substrate concentration, so the concentrations of the molecules have an influence on how the enzyme reacts.

REVISION A: As enzyme concentration and substrate concentration increased, so did the reaction rate. (Rewrite the sentence. The second half of the original sentence was deleted because it says nothing meaningful.)

REVISION B: Enzyme and substrate concentration influence enzyme reaction rate: an increase in enzyme or substrate concentration increased reaction rate. (Use a colon.)

Spelling

Spell checkers in word processing programs are so easy to use that there is really no excuse for *not* using them. Just remember that spell checkers may not know scientific terminology, so consult your textbook or laboratory manual for correct spelling. In some cases, the spell checker may even try to get you to change a properly used scientific word to an inappropriate word that happens to be in its database (for example, *absorbance* to *absorbency*).

The following poem is an example of how indiscriminate use of the spell checker can produce garbage:

Wrest a Spell

Eye halve a spelling chequer
It came with my pea sea
It plainly marques four my revue
Miss steaks eye kin knot sea.

Eye strike a key and type a word
And weight four it two say
Weather eye am wrong oar write
It shows me strait a weigh.

As soon as a mist ache is maid
It nose bee fore two long
And eye can put the error rite
Its rare lea ever wrong.

Eye have run this poem threw it
I am shore your pleased two no
Its letter perfect awl the weigh
My chequer tolled me sew.

— Sauce unknown

Spell checkers will also not catch mistakes of usage, for example *form* if you really meant *from*. Print out your document and proofread the hard copy carefully.

Punctuation

The purpose of punctuation marks is to divide sentences and parts of sentences to make the meaning clear. A few of the most common punctuation marks and their uses are described in this section. For a more comprehensive, but still concise, treatment of punctuation, see Hacker and Sommers (2016) or Lunsford (2013, 2015, 2016).

The comma The comma inserts a pause in the sentence in order to avoid confusion. Note the ambiguity in the following sentence:

> While the sample was incubating the students prepared the solutions for the experiment.

A comma *should* be used in the following situations:

1. To connect two independent clauses that are joined by *and, but, or, nor, for, so,* or *yet*. An independent clause contains a subject and a verb, and can stand alone as a sentence.

   EXAMPLE: Feel free to call me at home, but don't call after 9 p.m.

2. After an introductory clause, to separate the clause from the main body of the sentence.

   EXAMPLE: Although she spent many hours writing her lab report, she earned a low grade because she forgot to answer the questions in the laboratory exercise.

   A comma is not needed if the clause is short.

   EXAMPLE: Suddenly, the power went out.

3. Between items in a series, including the last two.

   EXAMPLE: Enzyme activity is affected by factors such as substrate concentration, pH, temperature, and salt.

4. Between coordinate adjectives (if the adjectives can be connected with *and*).

   EXAMPLE: The students' original, humorous remarks made my class today particularly enjoyable. (*Original and humorous remarks* makes sense.)

   A comma is not needed if the adjectives are cumulative (if the adjectives cannot be connected with *and*).

   EXAMPLE: The three tall students look like football players. (It would sound strange to say *three and tall students*.)

5. With *which*, but not *that* (see Word usage; that, which, p. 119)

6. To set off conjunctive adverbs such as *however, therefore, moreover, consequently, instead, likewise, nevertheless, similarly, subsequently, accordingly,* and *finally.*

   EXAMPLE: Instructors expect students to hand in their work on time; however, illness and personal emergencies are acceptable excuses.

7. To set off transitional expressions such as *for example, as a result, in conclusion, in other words, on the contrary,* and *on the other hand.*

   EXAMPLE: Chuck participates in many extracurricular activities in college. As a result, he rarely gets enough sleep.

8. To set off parenthetical expressions. Parenthetical expressions are statements that provide additional information; however, they interrupt the flow of the sentence.

   EXAMPLE: Fluency in a foreign language, as we all know, requires years of instruction and practice.

A comma *should not* be used in the following situations.

1. After *that*, when *that* is used in an introductory clause

   EXAMPLE: The student could not believe that he lost points on his laboratory report because of a few spelling mistakes.

2. Between cumulative adjectives, which are adjectives that would not make sense if separated by the word *and* (see Item 4 in preceding list)

The semicolon The semicolon inserts a stop between two independent clauses not joined by a coordinating conjunction (*and, but, or, nor, for, so,* or *yet*). Each independent clause (one that contains a subject and a verb) could stand alone as a sentence, but the semicolon indicates a closer relationship between the clauses than if they were written as separate sentences.

EXAMPLE: Outstanding student-athletes use their time wisely; this trait makes them highly sought after by employers.

A semicolon is also used to separate items in a series in which the items are already separated by commas.

EXAMPLE: Participating in sports has many advantages. First, you are doing something good for your health; second, you enjoy the camaraderie of people with a common interest; third, you learn discipline, which helps you make effective use of your time.

The colon The colon is used to call attention to the words that follow it. Some conventional uses of a colon are shown in the following examples.

Dear Sir or Madam:
5:30 P.M.
2:1 (ratio)

In references, to separate the place of publication and the publisher, as in

Sunderland (MA): Sinauer Associates, Inc.

A colon is often used to set off a list, as in the following example.

EXAMPLE: Catalase activity has been found in the following vegetables: turnips, leeks, parsnips, onions, zucchini, carrots, and broccoli.

A colon *should not* be used when the list follows the words *are, consist of, such as, including,* or *for example.*

EXAMPLE: Catalase activity has been found in vegetables such as turnips, leeks, parsnips, onions, zucchini, carrots, and broccoli.

The period The period is used to end all sentences except questions and exclamations. It is also used in American English for some abbreviations, for example, *Mr., Ms., Dr., Ph.D., i.e.,* and *e.g.*

Parentheses Parentheses are used mainly in two situations in scientific writing: to enclose supplemental material and to enclose references to visuals or sources. Use parentheses sparingly because they interrupt the flow of the sentence.

EXAMPLE:	Human error (failure to make the solutions correctly, arithmetic errors, and failure to zero the spectrophotometer) was the main reason for the unexpected results.
REFERENCE TO VISUAL:	There was no catalase activity above 70 °C (Figure 1).
CITATION-SEQUENCE SYSTEM:	C-fern spores do not germinate in the dark (1).
NAME-YEAR SYSTEM:	C-fern spores do not germinate in the dark (Cooke and others 1987).

The dash The dash is used to set off material that requires special emphasis. To make a dash on the computer, type two hyphens without a space before, after, or in between. In some word processing programs, the two hyphens are automatically converted to a solid dash.

Similar to commas and parentheses, a pair of dashes may be used to set off supplemental material.

EXAMPLE:	Human error--failure to make the solutions correctly, arithmetic errors, and failure to zero the spectrophotometer--was the main reason for the unexpected results. (If the word processing program has been set up to convert the two hyphens to a solid dash, the sentence looks like this: Human error—failure to make... spectrophotometer—was the main reason...)

Similar to a colon, a single dash calls attention to the information that follows it.

EXAMPLE:	Catalase activity has been found in many vegetables—turnips, leeks, parsnips, onions, zucchini, carrots, and broccoli.

If an abrupt or dramatic interruption is desired, use a dash. If the writing is more formal or the interruption should be less conspicuous, use one of the other three punctuation marks. However, do not replace a pair of dashes with commas when the material to be set off already contains commas, as in the following example.

EXAMPLE: The instruments that she plays—oboe, guitar, and piano—are not traditionally used in the marching band.

Abbreviations

The CSE Manual (2014) defines standard abbreviations for authors and publishers in the sciences and mathematics. Some of the terms and abbreviations that you may encounter in introductory biology courses are shown in Table 5.5. Take note of spacing, case (capital or lowercase letters), and punctuation use. Except where noted, the symbols are the same for singular and plural terms (for example, 30 min *not* 30 mins).

Widely known abbreviations such as DNA and ATP do not have to be defined. But abbreviations known only to specialists should be defined the first time they are used.

EXAMPLE: CRISPR (clustered regularly interspaced short palindromic repeats) technology makes it possible to edit segments of DNA in a precise and predictable fashion.

Numbers

Numbers are used for quantitative measurements. In the past, numbers less than 10 were spelled out, and larger numbers were written as numerals. The modern scientific number style recommended in the CSE Manual (2014) aims for a more consistent usage of numbers. The rules are as follows:

1. Use numerals to express any *quantity*. This form increases their visibility in scientific writing, and emphasizes their importance.
   - Cardinal numbers, for example, 3 observations, 5 samples, 2 times
   - In conjunction with a unit, for example, 5 g, 0.5 mm, 37 °C, 50%, 1 hr. Pay attention to spacing, capitalization, and punctuation (see Table 5.5).
   - Mathematical relationships, for example, 1:5 dilution, 1000× magnification, 10-fold

TABLE 5.5 Standard abbreviations in scientific writing

Term	Symbol or Abbreviation	Example
Latin words and phrases [The CSE Manual (2014) recommends that Latin words be replaced with English equivalents.]		[The Latin word may be replaced with the English equivalent given in brackets.]
circa (approximately)	ca.	The lake is ca. [approx.] 300 m deep.
et alii (and others)	et al.	Jones et al. [and others] (1999) found that . . .
et cetera (and so forth)	etc.	pH, alkalinity, etc. [and other characteristics] were measured.
exempli gratia (for example)	e.g.	Water quality characteristics (e.g., [for example,] pH, alkalinity) were measured.
id est (that is)	i.e.	The enzyme was denatured at high temperatures, i.e., the enzyme activity was zero. [Because the enzyme was denatured at high temperatures, the enzyme activity was zero.]
nota bene (take notice)	NB	NB [Important!]: Never add water to acid when making a solution.
LENGTH		
nanometer (10^{-9} meter)	nm	*Note:* There is a space between the number and the abbreviation. There is no period after the abbreviation.
micron (10^{-6} meter)	µm	
millimeter (10^{-3} meter)	mm	
centimeter (10^{-2} meter)	cm	
meter	m	450 nm, 10 µm, 2.5 cm
MASS		
nanogram (10^{-9} gram)	ng	*Note*: There is a space between the number and the abbreviation. There is no period after the abbreviation.
microgram (10^{-6} gram)	µg	
milligram (10^{-3} gram)	mg	
gram	g	
kilogram (10^{3} gram)	kg	450 ng, 100 µg, 2.5 g, 10 kg

TABLE 5.5 *(continued)*

Term	Symbol or Abbreviation	Example
VOLUME		
microliter (10^{-6} liter)	µl or µL	*Note*: There is a space between the number and the abbreviation. There is no period after the abbreviation.
milliliter (10^{-3} liter)	ml or mL	
liter	l or L	
cubic centimeter (ca. 1 mL)	cm^3	450 µl or 450 µL, 0.45 ml or 0.45 mL, 2 l or 2 L
TIME		
seconds	s or sec	*Note*: There is a space between the number and the abbreviation. There is no period after the abbreviation.
minutes	min	
hours	h or hr	
days	d	60 s or 60 sec, 60 min, 24 h or 24 hr, 1 d
CONCENTRATION		
molar (U.S. use)	M	TBS contains 0.01 M Tris-HCl, pH 7.4 and 0.15 M NaCl.
molar (SI units)	mol L^{-1}	
parts per thousand	ppt	Brine shrimp can be raised in 35 ppt seawater.
OTHER		
degree(s) Celsius	°C	15 °C (there is a space between number and symbol)
degree(s) Fahrenheit	°F	59 °F (there is a space between number and symbol)
diameter	diam.	pipe diam. was 10 cm
figure, figures	Fig., Figs.	As shown by Fig. 1, …
foot-candle	fc or ft-c	500 fc or 500 ft-c
maximum	max	The max enzyme activity was found at 36 °C.
minimum	min	The min temperature of hatching was 12 °C.
mole	mol	
percent	%	95% (there is no space between number and symbol)
species (sing.)	sp.	*Tetrahymena* sp.
species (plur.)	spp.	*Tetrahymena* spp.

2. Spell out numbers in the following cases:
   - When the number begins a sentence, for example, *"Twelve g of turnips was* (not *were*) *homogenized."* rather than *"12 g of turnips was homogenized."* Alternatively, restructure the sentence so that the number does not begin the sentence. Notice that when numbers are used in conjunction with units, the quantity is considered to be singular, not plural.
   - When there are two adjacent numbers, retain the numeral that goes with the unit, and spell out the other one. An example of this is *The solution was divided into four 250-mL flasks.*
   - When the number is used in a nonquantitative sense, for example, *one of the treatments, the expression approaches zero, one must consider...*
   - When the number is an ordinal number less than 10, and when the number expresses rank rather than quantity, for example, *the second time, was first discovered.*
   - When the number is a fraction used in running text, for example, *one-half of the homogenate, nearly three-quarters of the plants.* When the precise value of a fraction is required, however, use the decimal form, for example, *0.5 L* rather than *one-half liter.*
3. Use scientific notation for very large or very small numbers. For the number 5,000,000, write 5×10^6, not *5 million*. For the number *0.000005*, write 5×10^{-6}.
4. For decimal numbers less than one, always mark the ones column with a zero. For example, write *0.05*, not *.05*.

Format

Most university writing centers and professional editors recommend proofreading your paper in multiple "passes," looking for one kind of error in each pass (The Writing Center at UNC Chapel Hill 2014; CUNY Writing Fellows 2016; Every 2012). This strategy works particularly well for finding formatting errors, which are much easier to detect on printed pages than on the computer screen (Table 5.6). Check for potential errors in the following areas:

- Section headings
- Bulleted or numbered lists
- Figures and tables, including their in-text references
- In-text citations of outside sources
- Full references at the end of the paper

TABLE 5.6 Checklist for proofreading format

Category	Check for
Section headings	Correct order, consistent format, not separated from section body
Lists (bulleted or numbered)	Sequential numbering and consistent style, parallelism in sentence structure, consistent indentation for each level
Figures and tables	Sequential numbering in the order they are described
In-text references to figure and table numbers	Correspondence with the actual figures and tables
In-text citations	One-to-one correspondence with the end references; formatting is correct
End references	One-to-one correspondence with the in-text citations; all information is present; formatting is correct
Headers and footers (if needed)	Correct position on each page
Page numbers	Sequential (check especially after section breaks in Microsoft Word)
Typography	Consistent typeface, font size, and line spacing

- Headers and footers
- Page numbers
- Typography

Get Feedback

When we are engrossed in our work, we may fail to recognize that what is obvious to us is not obvious to an "outsider." That is where feedback from someone who is familiar with the subject matter comes in handy. If your instructor allows it, ask your lab partner, another classmate, or your teaching assistant to review your paper. Return the favor by reviewing someone else's. You may also get valuable feedback from a writing expert at your school's writing center.

The questions your reviewer will focus on are as follows:

- Do I know what the writer is trying to accomplish with this paper? Is the purpose clear?
- What questions or concerns do I have about this paper? Are there sections that were difficult to follow? Are the organization, content, flow, and level appropriate for the intended audience?

- What suggestions can I offer the writer to help him/her clarify the intended meaning?
- What do I like about the paper? What are its strengths?

Tips for being a good peer reviewer

There are two issues with which you may struggle when you are asked to review your classmate's paper: (1) I'm not confident that I know the "right" answer or know enough about the writing process to give good suggestions, and (2) I don't want to hurt the writer's feelings. These are valid concerns, and resolving them will require, first, a willingness to learn as much about writing scientific papers as possible, and second, the attitude that if something is unclear to you, it may also be unclear to other readers. With each paper you review, you will gain more confidence in your ability to give constructive feedback. In the meantime, however, a good rule of thumb is to give the kinds of suggestions and consideration that you would like to receive on your own paper.

When reviewing electronic files, the **New Comment** and **Track Changes** commands on the **Review** tab in Microsoft Word are very useful (see Appendix 1, pp. 199–201). **New Comment** allows the reviewer to make a comment or query, without editing the text itself, off to the side of the main text. When **Track Changes** is turned on, the reviewer's suggested changes, typed right into the body of the text, appear in a different color. The author can then accept or reject the changes. Think of the peer review process as a team sport: the reviewer is not challenging the writer's right to be on the team. The two are working together to get the best possible result.

Here are some concrete tips for being a good peer reviewer:

- Talk to the writer about his or her objectives, questions and concerns, parts that need specific feedback, and perceived strengths and weaknesses.
- Use the "Biology Lab Report Checklist" (pp. 138–140) for content.
- Look over Table 6.2 and "A Lab Report in Need of Revision" (pp. 150–161 in Chapter 6) for common errors.
- Mark awkward sentences, spelling and punctuation mistakes, and formatting errors. Do not feel you have to rewrite individual sentences—that is the writer's job.
- Ask questions. Let the writer know where you can't follow his/her thinking, where you need more examples, where you expect more detailed analysis, and so on.

- Do not be embarrassed about making lots of comments; the author does not have to accept your suggestions. On the other hand, if you say only good things about the paper, how will the writer know whether the paper is accomplishing the desired objectives?

You can fine-tune your proofreading skills on any text. You may recognize some of your own problems in other people's writing, and, with persistence and practice, you will find creative solutions to correct these problems. **Keep a log of the problems that recur in your writing** and review them from time to time. Repetition builds awareness, which will help you achieve greater clarity in your writing.

Have an informal discussion with your peer reviewer

Sometimes the comments made by the peer reviewer are self-explanatory. Other times, however, the peer reviewer cannot respond to certain parts of the paper, because more information is required. Under these circumstances, an informal discussion between the writer and the reviewer is helpful. There are two important rules for this discussion:

- First, the writer talks and the reviewer listens. The objective is to help the writer express exactly what he/she wants to say in the paper.
- Second, the reviewer talks, in nonjudgmental terms, about which parts of the paper were readily understandable and which parts were confusing. The reviewer does not have to be an experienced writer to do this—no two people have exactly the same life experiences, and there is always something positive you can learn from looking at your writing from someone else's perspective.

Feedback from your instructor

Some instructors write comments on the hard copy of a lab report; others use electronic editing tools. When instructors grade hard copies, they usually use standard proofreading marks to save time. If you don't know what the marks mean, ask your instructor or look them up. Frequently used proofreading marks are listed in Table 6.1. A more comprehensive list is available in the CSE Manual (2014), at http://www.biomedicaleditor.com/support-files/proofreadingmarks.pdf, and in other printed and online sources.

Online submissions have become increasingly widespread, especially at the college level, for several reasons. Students like the convenience. Submitting assignments online does not require a printer and can be done at the last minute. When a document is stored in the cloud, it can be accessed from a variety of devices, and there is also less chance that it will get lost (physically, at least). Going paperless is also good for the environment.

Online submissions have benefits for instructors as well, which include:

- Being able to provide higher quality, more consistent feedback in less time.
- Having a time stamp to confirm when a student's assignment was turned in.
- Having the capability to check for plagiarism automatically.

Most of the concerns that students and instructors voice about online submissions are related to technical difficulties during the upload process, changes in document format, the inability to resubmit a document if an error is discovered after the fact, unreliable Internet connections, and a personal dislike of technology in general. Ultimately, it is up to your instructor to decide whether the advantages of online submission outweigh the disadvantages.

Instructors have various options for providing feedback in your electronic documents. One option is to insert comments or mark up text using the Comments and/or Track Changes features in Word (see pp. 199–201 in Appendix 1). A second option that applies specifically to PDFs is to open the PDF in Adobe Acrobat Reader. In this program, text can be highlighted and comments can be added on "sticky notes." A third option for providing feedback can be used on either Word documents or PDFs uploaded to Turnitin. In Turnitin Classic, instructors can highlight text on the paper, click the **Comment** button on the side panel, and then type a comment (Figure 5.1, item ①). If this turns out to be a frequently made comment, it can be saved as a GradeMark and reused on the current paper or on other students' papers. The title of the GradeMark is limited to 40 characters, but there is plenty of room to add detailed explanations in the context box. The next time this comment is applicable, instructors simply drag the GradeMark off the side panel and drop it onto the student's paper (Figure 5.1, item ②). Additional comments can be added to existing GradeMarks; these additional comments are only displayed in that particular comment in that particular paper. However, universal edits to existing GradeMarks will be applied to all papers in which that GradeMark was used. Two other features that instructors will appreciate are

- GradeMarks can be saved in sets customized for individual assignments (Figure 5.1, item ③) and
- GradeMark sets can be shared, which allows grading to be more consistent across multiple lab sections.

Students are able to see their instructor's comments by clicking the assignment link to open their paper, clicking the large GradeMark but-

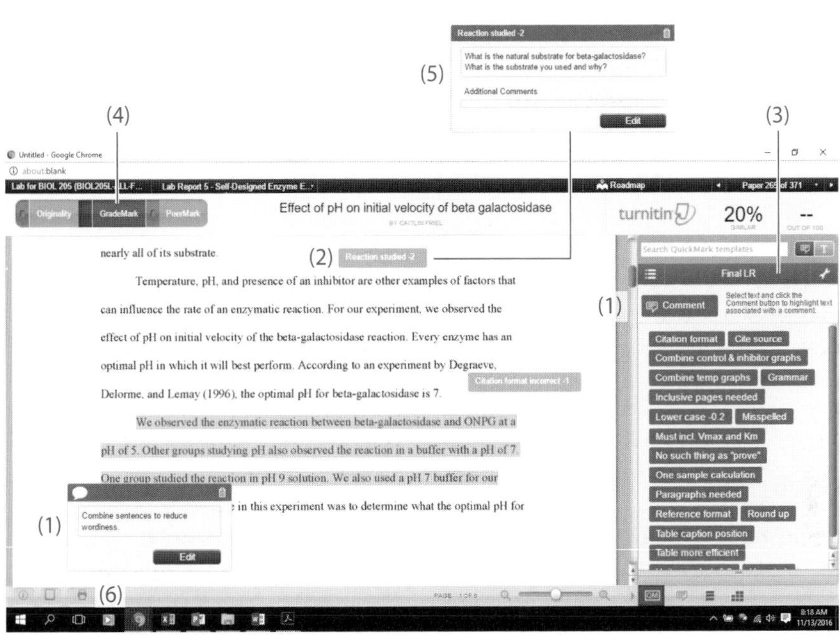

Figure 5.1 Comments and GradeMarks in Turnitin make it possible to provide consistent and detailed feedback on assignments. (1) After highlighting text, click Comment on the side panel to insert a comment that will only be displayed on the current paper. (2) Comments saved as GradeMarks can be reused in all assignments. (3) GradeMarks are organized into sets that can be customized for individual assignments. (4) Students click the GradeMark button to see their instructor's comments. (5) Comments expand when the cursor is held over the comment. (6) Assignments with comments can be downloaded as PDFs or printed for future reference.

ton in the top left corner of the screen (Figure 5.1, item ④), and then holding the cursor over each comment (Figure 5.1, item ⑤). The document along with all of the expanded comments can be downloaded as a PDF or printed out for future reference (Figure 5.1, item ⑥). In the new version of Turnitin called *Feedback Studio,* some of these functions have been improved. For a comparison of Turnitin Classic and Feedback Studio, see, for example, https://www.youtube.com/watch?v= tIKjBzJIe2g.

Biology Lab Report Checklist

TITLE (pp. 88–89)

- ☐ Descriptive and concise

AUTHORS

- ☐ Each author's first name is followed by his/her surname

ABSTRACT (pp. 87–88)

- ☐ Contains an introduction (background and objectives)
- ☐ Contains brief description of methods
- ☐ Contains results
- ☐ Contains conclusions

INTRODUCTION

- ☐ Starts with a general introduction to the topic (pp. 84–86)
- ☐ Contains a question or unresolved problem (pp. 84–86)
- ☐ Contains background information supported by in-text references (pp. 84–86)
- ☐ The selected references are directly relevant to the study (pp. 84–86).
- ☐ The in-text reference is formatted correctly according to the Name-Year, Citation-Sequence, or Citation-Name system (pp. 89–95).
- ☐ Information obtained from a reference is paraphrased. Direct quotations are not used (p. 90).
- ☐ The objectives of the study are clearly stated (pp. 84–86).

MATERIALS AND METHODS

- ☐ Contains all relevant information to enable a person with appropriate training to repeat the procedure (pp. 55–57).
- ☐ Routine procedures are not explained (pp. 57–58).
- ☐ Complete sentences and paragraphs are used—do not make a numbered list (p. 55).
- ☐ Past tense is used because these actions were done in the past and completed (p. 55).
- ☐ Passive voice is used to emphasize the action (active voice is allowed in some disciplines) (p. 55).

- ☐ Materials are not listed separately (p. 57).
- ☐ No preview is given of how the data will be graphed or tabulated (p. 58).

RESULTS

- ☐ Tables and figures are described in numerical order. The descriptive text for a table or figure immediately precedes that table or figure (see, for example, Figures 4.1, 4.3, and 4.4).
- ☐ Results are described in *past* tense (p. 79).
- ☐ Every sentence in the text is meaningful (p. 79 and pp. 108–110). Sentences such as *The results are shown in the figure below* are not meaningful.
- ☐ When a result is described, the figure showing that result is referenced, preferably in parentheses at the end of the sentence (p. 77–78).
- ☐ There are no tables and figures that are not described.
- ☐ The figure caption is positioned *below* the figure (p. 67). The table caption is positioned *above* the table (p. 63).
- ☐ Figure and table titles are informative and can be understood apart from the text (pp. 63–64 and 67–69).
- ☐ The results are not explained (p. 78).

DISCUSSION (pp. 81–84)

- ☐ The results are *briefly* restated.
- ☐ The results are explained and interpreted.
- ☐ Past tense is used when referring to your own results. Present tense is used to state scientific fact, which is information supported by experimental evidence and replicated by many different scientists. Results in journal articles are considered to be fact until other studies present evidence to the contrary (p. 54).
- ☐ The results are compared with those in journal articles.
- ☐ The results are related back to the original objectives stated in the Introduction.
- ☐ Any errors and inconsistencies may be pointed out.
- ☐ The significance of the results or their implications may be discussed in a broader context.

REFERENCES (pp. 89–101)

- ☐ The references consist mostly of journal articles, not secondary sources such as textbooks or websites.
- ☐ The references are formatted correctly and contain all the required information.
- ☐ All references listed in this section have been cited in the text. All in-text references have been included in this section.
- ☐ Reference management software saves time formatting references (pp. 24–30).

REVISION

- ☐ All questions from the laboratory exercise have been answered.
- ☐ Calculations and statistics have been double-checked (p. 105).
- ☐ The overall structure of the manuscript is correct (pp. 104–105).
- ☐ The overall structure of each section is correct (pp. 105–106).
- ☐ Figures and tables are formatted correctly (pp. 60–73).
- ☐ Sections, paragraphs, sentences, and words are coherent and meaningful (pp. 105–112).
- ☐ Individual words are used appropriately for the situation (pp. 112–120).
- ☐ All sentences are grammatically correct (pp. 121–124).
- ☐ All words are spelled correctly (p. 124).
- ☐ The correct punctuation marks are used (pp. 125–129).
- ☐ Abbreviations for unfamiliar terms are defined the first time they are used (p. 129).
- ☐ Standard abbreviations are used for units (Table 5.5; pp. 130–131).
- ☐ Numbers are formatted correctly and, when applicable, are followed by units (pp. 129–132).
- ☐ The format for section headings, lists, figures, tables, references, and typography is consistent (Table 5.6; pp. 132–133).

Go to the **COMPANION WEBSITE** • **sites.sinauer.com/knisely5e**
for samples, template files, and tutorial videos

Chapter 6

Sample Student Laboratory Reports

A "Good" Sample Student Lab Report

The first laboratory report in this chapter was written by Lynne Waldman during her first year at Bucknell University, in an introductory course for biology majors. Lynne and her lab partners designed and carried out an original project in which they investigated the effect of a fungus on the growth of bean, pea, and corn plants.

Lynne's report has many of the characteristics of a well-written scientific paper. When you look over her presentation, notice the style and tone of her writing, as well as the format of the paper. The comments and annotations in the margins alert you to important points to keep in mind when you write your laboratory report.

The presentation here has been typeset to fit this book and to accommodate the annotations. Your report should be formatted to fit standard 8.5 × 11 inch paper. Unless you are instructed otherwise, use a serif type (Times Roman is standard), double space, and leave at least 1 inch of margin all around.

For details on how to format documents in Microsoft Word, see "Formatting Documents" in Appendix 1.

Title is informative.

Write author's name first followed by lab partners' names.

The Effects of the Fungus *Phytophthora infestans* on Bean, Pea, and Corn Plants

Lynne Waldman, Partner One, Partner Two

Label sections of lab report clearly.

Provide background information.

State purpose of current experiment.

Briefly describe methods.

Do not cite sources in abstract.

Do not refer to any figures.

Describe results.

Briefly explain the results or state your conclusions.

Limit abstract to a maximum of 250 words.

Abstract

Phytophthora infestans is a fast-spreading, parasitic fungus that caused the infamous potato blight by devastating Ireland's crops in the 1840s. *P. infestans* also causes late blight in tomato plants, a relative of the potato. In this experiment, the effects of *P. infestans* on *Phaseolus* variety long bush bean, *Zea mays* (corn), and *Pisum sativum* (pea) were studied. The soil surrounding the roots of 18-day old plants was injected with *P. infestans* cultured in an L-broth medium. Plant height, number of leaves, and leaf angle were measured for each plant during the next 8 days. Chlorophyll assays were performed prior to exposure, and on the eighth day after exposure to the fungus. The plants were also examined for black or brown leaf spots characteristic of late blight infections. The results showed that *P. infestans* had no apparent effect on the bean, corn, and pea plants. One reason for this may be that there were no fungus zoospores in the L-broth medium. More probably, however, *P. infestans* may be a species-specific pathogen that cannot infect bean, corn, or pea plants.

Italicize Latin names.

Provide background information.

Introduction

Originating in Peruvian-Bolivian Andes, the potato (*Solanum tuberosum*) is one of the world's four most important food crops (along with wheat, rice, and corn). Cultivation of potatoes began in South America over 1,800 years ago, and through the Spanish conquistadors, the tuber was introduced into Europe in the second half of the 1600s. By the beginning of the 18th century, the potato was widely grown in Ireland, and the country's economy heavily relied on the potato crop. In the middle of the 19th century, Ireland's potato crop suffered widespread late blight disease caused by *Phytophthora infestans,* a species of pathogenic plant fungus. Failure of the potato crop because of late blight resulted in the Irish potato famine. The famine led to widespread starvation and the death of about a million Irish people.

The potato continues to be one of the world's main food crops. However, *P. infestans* has reemerged in a chemical-resistant form in the United States, Canada, Mexico, and Europe (McElreath, 1994). Late blight caused by the new strains is costing growers worldwide about $3 billion annually. The need to apply chemical fungicides eight to ten times a season further increases the cost to the grower (Stanley, 1994 and Stanley, 1997). *P. infestans* is thus an economically important pathogen.

Use proper citation format (e.g., Name-Year system).

Do not use direct quotations. Paraphrase source text and cite the source in parentheses.

P. infestans, which can destroy a potato crop in the field or in storage, thrives in warm, damp weather. The parasitic fungus causes black or purple lesions on a potato plant's stem and leaves. As a result of infection by this fungus, the plant is unable to photosynthesize, develops a slimy rot, and dies. *P. infestans* similarly infects the tomato plant (*Lycopersicon esculentum*) (Brave New Potato, 1994).

Use an abbreviated title when no author is given.

The purpose of the present experiment was to determine the effects of *P. infestans* on plant height, number of leaves, leaf angle, and chlorophyll content of three agriculturally important plants: *Phaseolus* variety long bush beans, *Zea mays* (corn), and *Pisum sativum* (peas). Symptoms of fungal infection were assumed to be similar to that in potatoes.

State purpose of experiment clearly.

Materials and Methods

Phaseolus variety long bush bean, *Zea mays* (corn), and *Pisum sativum* (pea) seeds were soaked overnight in tap water. Fifteen randomly chosen seeds of each species were planted 1 cm beneath the surface in three separate trays containing 10 cm of potting soil. Another set of trays, which was to be the control group, was prepared in the same fashion. All the experimental plants were placed in one fume hood, and all the control plants were placed in relative positions in another fume hood in the same room. The plants were exposed to the ambient light intensity in the hood (153 fc) and air current 24 hrs a day, and were

Write Materials and Methods section in past tense.

Provide sufficient detail to allow the reader to repeat the experiment.

watered lightly daily. The plants were allowed to germinate and grow for 18 days.

Phytophthora infestans on potato dextrose agar was obtained from Carolina Biological Supply House. At day 10 of the plant growth regime, pieces of agar on which the fungus was growing were transferred to L-broth. L-broth consisted of 5 g yeast extract, 10 g tryptone, 1 g dextrose, and 10 g NaCl dissolved in distilled water, and adjusted to pH 7.1, to make 1 L of medium. The medium was sterilized before adding the fungal culture. After 4 days in L-broth, 6 mL of the fungal culture was injected into the soil around the roots of each 18-day old plant. Six mL of L-broth without *P. infestans* was injected into the soil of the control plants. All plants were then allowed to grow for another 8 days.

Every other day after treatment with *P. infestans*, plant height and number of leaves were measured for both the control and the experimental plants. Plant height was measured from the soil to the apical meristem of the plant. Leaf angle (as shown in Figure 1) of the largest, lowest leaf on each plant was measured three times, once prior to injection, once 4 days after injection, and once 8 days after injection. Leaf angle was measured in order to determine if *P. infestans* causes wilting in the three plant species. In addition, the plant was examined visually for the presence of any leaf spots.

Chlorophyll assays were performed on one plant from each tray prior to injection and on the eighth day after injection. For each chlorophyll assay, the leaves of the plant were removed from the stem. For each 0.1 g of

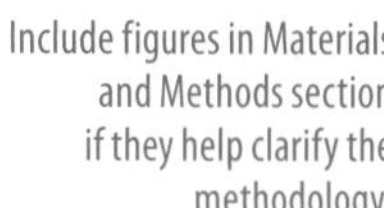
Include figures in Materials and Methods section if they help clarify the methodology.

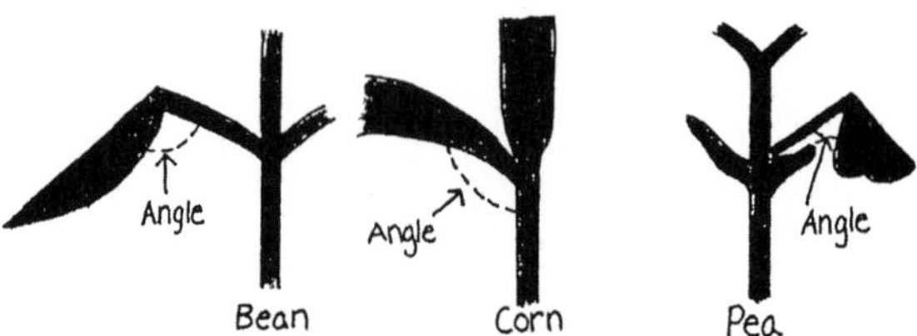

Figure 1 Leaf angle as measured in bean, corn, and pea plants

leaves, 6.0 mL of 100% methanol were used. The leaves were thoroughly ground in half of the methanol with a pestle in a mortar. The leaves were ground again after the rest of the methanol was added. Extraction of the chlorophyll was allowed to proceed for 45 min at room temperature. Then the suspension was gravity filtered through filter paper to remove the leaf parts. The absorbance of the filtrate was measured with a Spectronic 20 spectrophotometer at 652 nm and 665.2 nm. The absorbance values were converted to relative chlorophyll units using the following equation derived by Porra and colleagues (1989):

Total chlorophyll (a and b) = Dilution factor × $[22.12\ A_{652\ nm} + 2.71 A_{665.2\ nm}$ (mg/L)] × Volume of solvent (L) / Weight of leaves (mg)

Make proper subscripts.

Results

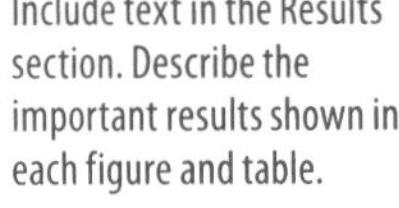

Include text in the Results section. Describe the important results shown in each figure and table.

P. infestans-treated plants and the control plants had similar growth patterns (Figure 2). Both the experimental and control pea and corn plants grew at a constant, but very slow rate over the eight day test period. The control bean plants were taller on average than the experimental bean plants throughout most of the experiment. Both groups showed the same growth pattern, however, with rapid growth occurring from day 18 to 24 (0 to 4 days after injection), followed by slower growth to the end of the experiment.

As plant height increased, the average number of leaves on all of the plants also increased over the measurement period (Figure 3). There is an

Refer to each figure and table in parentheses.

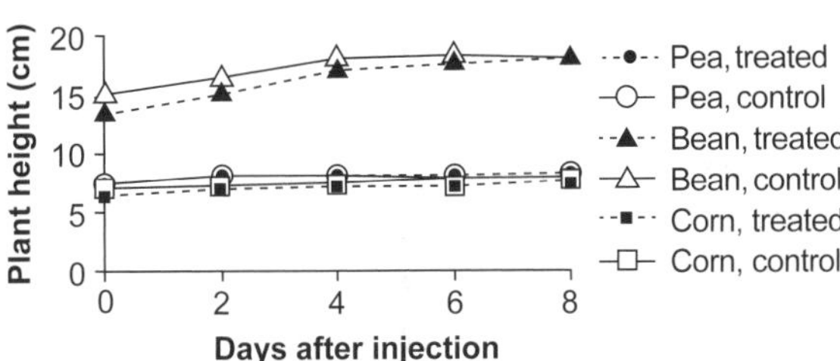

Make text in legend and in axes titles large enough to read easily.

Make sure intervals on axes have correct spacing.

Figure 2 Average height of control and experimental plants in the period after injection with *P. infestans*

Make points and lines black and background white for best contrast.

In the axes titles, write the variable followed by the units in parentheses (where applicable).

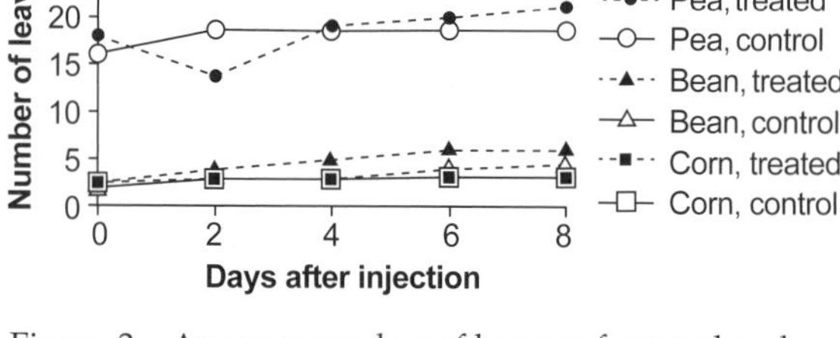

Position the figure caption below the figure.

Figure 3 Average number of leaves of control and experimental plants in the period after injection with *P. infestans*

uncharacteristic decrease in the number of leaves of pea plants treated with *P. infestans* from day 18 to 20 (0 to 2 days after injection), but this is probably due to counting error.

Describe the figures in order.

Insert symbols such as ° using word processing software.

There was a general decline in average leaf angle of all the plants over the first four days after injection with *P. infestans* (Figure 4). The plants did not follow this pattern over the second half of the experiment, however. The leaf angle of the experimental bean group increased by 28°, while that of the control bean group only increased by about 3°. The leaf angle of the control pea plants increased significantly (33°), while that of the experimental pea plants decreased 4°. The leaf angle of the corn control group decreased 0.5°, while that of the corn experimental group showed a much sharper decline of 24°.

There was also no difference between the experimental and control groups with regard to

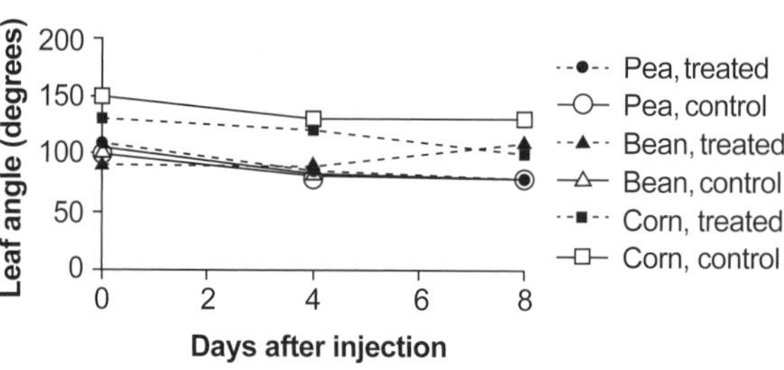

Make the figure title descriptive. Do not use "y-axis title vs. x-axis title."

Figure 4 Average leaf angle of control and experimental plants in the period after injection with *P. infestans*

Table 1 Chlorophyll content of corn, bean, and pea plants prior to infection and 8 days after infection

Position the table caption above the table.

Plants	Relative chlorophyll units		Change in chlorophyll content (relative units)
	Day 0	Day 8	
Corn, treated	9.036×10^{-4}	9.383×10^{-4}	$+3.45 \times 10^{-5}$
Corn, control	9.270×10^{-4}	8.963×10^{-4}	$+3.34 \times 10^{-5}$
Bean, treated	1.034×10^{-1}	1.2×10^{-3}	-1.022×10^{-1}
Bean, control	1.7×10^{-3}	1.6×10^{-3}	-1×10^{-4}
Pea, treated	1.3×10^{-3}	1.7×10^{-3}	$+4 \times 10^{-4}$
Pea, control	1.2×10^{-3}	1.2×10^{-3}	0.0000

Use scientific notation correctly.

Do not split small tables across two pages.

chlorophyll content. There was a slight increase in chlorophyll content from day 18 to 26 (0 to 8 days after injection) in the corn plants (Table 1). For the bean group, there was a large decrease in chlorophyll content, 0.1 relative chlorophyll units, which did not seem to agree with the general appearance of the plants. There may have been some error when this assay was carried out. There was little change in chlorophyll content for the pea group.

Number tables independently of figures.

Finally, there was no evidence of any brown or black leaf spots symptomatic of *P. infestans* infection.

Discussion

P. infestans did not affect the plant height, leaf angle, number of leaves, and chlorophyll content of *Zea mays, Pisum sativum*, or *Phaseolus*. Symptoms of infection are the presence of brown or black spots (areas of dead tissue) on leaves and stems, and, as the infection spreads, the entire plant becomes covered with a cottony film (Stanley, 1994). None of the experimental plants exhibited these symptoms.

Briefly restate the results in the Discussion section.

There may be several reasons why *P. infestans* did not affect the plants in this study. One reason is that the L-broth culture of *P. infestans* may not have contained zoospores of the fungus. Zoospores are motile spores that can penetrate the host plant through the leaves and soft shoots, or through the roots (Stanley, 1994). Zoospores are usually produced in wet,

Give possible explanations for the results.

Support explanations with references to published sources.

warm weather conditions (Ingold and Hudson, 1993). If the L-broth culture did not contain any zoospores, or if the soil around the plants was not sufficiently saturated to stimulate production of zoospores, then these conditions may have prevented *P. infestans* from attacking the roots and shoots of the plants.

Offer possible ways to test whether explanation is valid.

In order to determine if the problem was lack of zoospores, first the L-broth culture could be examined microscopically for presence of zoospores. Second, the *P. infestans* plants could be watered with different quantities of water to determine if the fungus requires wetter soil for zoospore production and motility.

Another reason why *P. infestans* may not have affected the plants is that this species of fungus may be specific to potato (*Solanum tuberosum*) and tomato (*Lycopersicon esculentum*) plants (Stanley, 1994), which both belong to the nightshade family (Solanaceae). In contrast, corn belongs to the grass family (Gramineae), and peas and beans are legumes (Leguminosae). It may be that these plant families are not susceptible to *P. infestans*, which has a very limited host range (Stanley, 1994). Non-susceptible plants have been shown to have defense mechanisms that prevent *P. infestans* from infecting them (Gallegly, 1995).

Whenever possible, use primary references (journal articles, conference proceedings, collections of primary articles in a book). Avoid unreliable websites.

Further research is required to determine if *P. infestans* really cannot infect corn, pea, and bean plants. Goth and Keane (1997) developed a test to measure resistance of potato and tomato varieties to original and new strains of *P. infestans*. Their experiments involved exposing the experimental plants' leaves directly to the fungus, and this method could perhaps be tested on corn, pea, and bean leaves as well.

References

Substitute the title when no author is given.

In Name-Year end reference format, list authors alphabetically by first author's last name.

Brave New Potato. 1994. Discover 15(10): 18–20.

Gallegly ME. 1995. New criteria for classifying Phytophthora and critique of existing approaches. In: Erwin DC, Bartnicki-Garcia S, Tsao PH, editors. Phytophthora: Its Biology, Taxonomy, Ecology, and Pathology St. Paul: The American Phytopathological Society. pp. 167–172.

Goth RW, Keane J. 1997. A detached-leaf method to evaluate late blight resistance in potato and tomato. American Potato Journal 74(5): 347–352.
Ingold CT, Hudson HJ. 1993. The Biology of Fungi, 6th ed. London: Chapman and Hall.
McElreath, Linda R. 1994. One potato, two potato. Agricultural Research 42(5): 2–3.
Porra RJ, Thompson WA, Kriedemann PE. 1989. Determination of accurate extinction coefficients and simultaneous equations for assaying chlorophylls a and b extracted with four different solvents: verification of the concentration of chlorophyll standards by atomic absorption spectroscopy. Biochimica et Biophysica Acta 975: 384–394.
Stanley D. 1994. What was around comes around. Agricultural Research 42(5): 4–8.
Stanley D. 1997. Potatoes. Agricultural Research 45(5): 10–14.

Use mostly primary journal articles or articles in a book.

List all authors (up to 10; then list first 10 followed by *et al.* or *and others*).

See the tabbed pages in Chapter 4 for examples of how to reference printed and electronic sources.

Give inclusive page numbers, not just the page(s) you extracted information from.

Make sure all in-text citations have a corresponding end reference.

Make sure all end references have a corresponding in-text citation.

Laboratory Report Errors

Table 6.1 lists proofreading marks that your instructor may use to give you feedback on your lab report. Marks are placed in the margin as well as in the text itself to describe the required revision.

Table 6.2 lists some common errors made by students writing biology lab reports. As instructors, we want students to be aware of these and other errors and to understand how to correct them. However, providing detailed explanations takes time, and time is a precious commodity. To save time without compromising quality, some instructors use tools for editing electronic documents, such as Word's Comments and Track Changes features and Turnitin's GradeMarks. Other instructors prefer to write abbreviated comments in the margin of printed documents. Either way, as a student, it is your responsibility to take your instructor's comments seriously and take the necessary steps to avoid making the same kinds of errors in the future. Tables 6.1 and 6.2 can be downloaded from <http://sites.sinauer.com/Knisely5E>. Instructors can edit the tables to customize the abbreviations for specific assignments. Students are expected to look up the meaning of each abbreviated comment for continuous improvement of their writing skills.

TABLE 6.1 A short list of proofreading marks

In Margin	In Text	Explanation
[	[More starch	Left-align text
]	(1)]	Right-align text
][	]Lab report title[	Center text
lc	PH	Use lowercase letter
CAP or uc	dna	Use capital (uppercase) letter
^to	refer ^ each	Insert text
ɣ	absolutely essential	Delete text
#	in|spite of	Insert space
⁀‿	to morrow	Close up
¶	starch. ¶The objective	Start a new paragraph
……	at room temperature	A dotted underline means "stet," or "let original text stand." The correction was made in error.
ital	E. coli	Italicize Latin names of organisms
V	6.02 × 10E23	Delete E and superscript exponent
^	Vmax	Subscript needed

TABLE 6.2 Abbreviations for comments made on lab reports

Abbreviation	Explanation
abbr	Write out the abbreviation the first time it is used, e.g., wild type (WT).
agr	Make subject and verb agree.
awk/incomplete	Revise to make less awkward. Convert sentence fragment to a complete sentence.
bkgd	Background is insufficient. Add details.
calc guid	Use full sentences to guide the reader through the calculation procedure. Use past tense (because the procedure was done in the past) and passive voice (to emphasize what was done, not who did it). For example, *The ___ was calculated using ___*. Show the original equation and define the terms. Then substitute known (or measured) values and solve for the unknown. State the final answer and include units.
caption pos	Figure caption goes *below* the figure. Table caption goes *above* the table. When you make the figure in Excel, leave the "Chart title" space blank.

TABLE 6.2 Abbreviations (*continued*)

Abbreviation	Explanation
cit form N-Y	• For 2 authors, include both authors' surnames separated by *and* followed by year of publication. • For 3 authors, include first author's surname followed by *and others* and year of publication.
cit form	Use CSE Name-Year, Citation-Sequence, or Citation-Name format. See Chapter 4 for specific examples. Do not use direct quotations. Paraphrase and cite the source.
cit missing	Cite all end references in the text.
command	Do not use command style. Reword in past tense. For example, rather than *Substitute the absorbance for y and solve for x*, write *The absorbance was substituted for y, and the equation was rearranged to solve for x*.
compare	• Compare treatment and control groups. • Compare your results with those in journal article.
content	Section is missing essential content
details	• The title is not descriptive. Add details such as variables and organism(s). • Essential details are missing in the Materials and Methods section. Provide enough detail to enable a trained person to repeat the experiment.
don't preview	In the Materials and Methods section, do not give a "preview" of how data will be plotted or tabulated in the Results section.
eq ed	Use Equation Editor in Word to make professional-looking equations. See Appendix 1.
expl	Explanation is insufficient.
fig format	Figure format is incorrect. See Appendix 2 to format graphs in Excel. • Delete gridlines and chart border. • Use CSE-preferred symbols: filled or open circles, squares, and triangles. • Make all lines and symbols black for best contrast. • Use outside tick marks. • On the axis label, put units in parentheses after the variable. • Use standard intervals in multiples of 2 or 5 on the axes. • Shorten axis to eliminate empty space.

TABLE 6.2 Abbreviations (*continued*)

Abbreviation	Explanation

	• Legend is not needed when there is only one data set.
	• Legend is needed to distinguish multiple data sets on one graph.
	  ∘ Move legend inside axes.
	  ∘ Make legend entries meaningful. See "Multiple Lines on an XY Graph" in Appendix 2.
	• Insert a line to show the trend. See "Choose a line" in Appendix 2 to decide which type of line.

fig/tab pos	Position the figure/table immediately after the text where it is first described. That way the reader will read the description first and know which visual contains the data.
fig/tab ref	Reference figure/table number in parentheses at the end of the sentence. Put the period after the closing parenthesis.
fig/tab title	• Figure/table title is factually incorrect. • Figure/table title is inadequate. Add details to make title self-explanatory. • Use sentence case (do not capitalize common nouns unless they start a sentence).
gram	Grammatical error
head-body sep	Keep section heading and body together. • **Windows:** On the ribbon, **Home | Paragraph diagonal arrow | Line and Page Breaks** tab. Check the **Keep with next** checkbox. • **Mac:** On the menu bar, **Format | Paragraph | Line and Page Breaks**. Check the **Keep with next** checkbox.
heading	Add section heading.
hyp	State the hypothesis.
interp	Interpret the results in the Discussion (not the Results) section.
math	Calculation error
meaning	Make sentences meaningful.
not a recipe	Do not list materials separately. Do not make a numbered or bulleted list. Use full sentences, paragraphs, and past tense to describe the procedure.
num format	• Put a zero in the ones place: 0.1 mL not .1 mL • Use scientific notation when numbers are very large or very small.

TABLE 6.2 Abbreviations (*continued*)

Abbreviation	Explanation
obj?	State the objective(s) of the current study.
page break	End the current page; move subsequent text to next page. • **Windows: Ctrl+Enter** • **Mac: ⌘+Enter**
¶	Break this section into paragraphs. When you start a new topic, start a new paragraph.
passive voice	Passive voice is preferred. Shift the emphasis from yourself to the subject of the action.
past tense	• Use past tense to state the objectives. • Use past tense to describe the procedure. • Use past tense to indicate that you are referring to your own results and not making a statement that is universally true.
present tense	Use present tense for scientific fact (information already accepted by the scientific community).
punc	Punctuation error
ref missing	List all in-text citations in the References section.
ref format	Reference elements are out of order. Information is missing.
rep	Eliminate repetition.
result?	• Describe the result, and reference the figure/table where the data are located. • Use specific language. How did the independent variable affect the dependent variable? What was the general relationship or trend?
round up	Round up final answer. The final answer cannot be more precise than the least precise measured value.
routine	Do not describe routine laboratory procedures in detail.
run-on	Break run-on sentence into two sentences.
scatter not line	Choose "XY Scatter" not "Line" in Excel to space data correctly. See Appendix 2.
source?	Cite sources to provide background information, to substantiate claims, and to compare findings.
sp	Word is misspelled
sub/super	• Superscript exponents. • Use AutoCorrect to format expressions with sub/superscripts automatically. See Appendix 1.

TABLE 6.2 Abbreviations (*continued*)

Abbreviation	Explanation
symbol	Choose the correct symbol from **Insert | Symbol** on the Ribbon. For example, insert °C instead of writing out *degrees Celsius*. See Appendix 1 for making shortcut keys.
tab unnec	Do not include a table when the graph shows the same data.
units	Units are missing or incorrect.
unnec intro	Do not write unnecessary introductory phrases like this. See Chapter 4.
wc	Word choice. This word does not fit the context. See Chapter 5.
wordy	Revise to reduce wordiness.

A Lab Report in Need of Revision

The following lab report contains the kinds of errors that are made by students who are just learning to write biology lab reports. Some of these errors have to do with writing in general (punctuation, word usage, sentence structure, transitions, coherent paragraphs, and so on) and others are specific to writing journal articles in biology. After writing but before revising the first draft of your lab report, skim the "lab report in need of revision" to anticipate the kinds of errors often made in the various sections. Use Tables 6.1 and 6.2 to look up the meaning of the abbreviations in the margin. Although some of these errors may not affect the content of your paper, they do affect your credibility as a scientist; careless writing may be equated with careless science.

Margin notes	Lab report
Details: What aspect of barley seeds did you study?	]Barley seed lab report[
Center title and author Delete *by*	]~~by~~ Ima Sprout[
	Abstract
abbr: ^gibberellic acid (GA) / wc: ^affects / ^include common name barley ital. Latin names Passive voice: use for procedure Content: results and disc missing	We looked at how ^(GA)~~effects~~ ^ the production of α-amylase in ^ Hordeum vulgare seeds. We cut seeds in half and gave them either water or a concentration of GA. After a week we measured how much starch was left in the seed. Our results showed that there was a correlation between GA concentration, how much starch was remaining, and the amount of α-amylase produced in the seeds.

Introduction

The concentration of GA was experimented on to see how this factor effected production of α-amylase. When a barley seed germinates, GA is released from the embryo. GA moves to the aleurone layer of the seed, where it activates the production of α-amylase (Tanaka). After its release, α-amylase diffuses through the endosperm and hydrolyzes starch into α-1,4-linked glucose oligosaccharides that are then broken down by other enzymes into maltose and glucose (Koning). There should thus be a correlation between the amount of GA and the digestion of the endosperm in barley seeds.

Our hypothesis was that higher GA concentrations would cause increased α-amylase production, which would cause a decrease in starch in the barley seed endosperm.

move obj to the end of the intro and make sentence less wordy; wc: affected not effected

define abbr

cit form N-Y

cit form N-Y

Materials and Methods

~~Obtain filter paper, forceps, a razor blade, 3 Petri dishes, barley seeds, 10 μM GA, 10 mM HEPES-EGTA-Ca2+, a centrifuge, starch solution, HCl, 10 mM citric acid-sodium citrate buffer, Lugol's iodine, and a spectrophotometer. The tests had either~~ W~~w~~hole barley seeds (Plate A) or half barley seeds (Plates B-F). ~~The half seeds had~~ ^ the embryo removed ~~and the~~ ^ cut ~~endosperm was placed~~ ^ down on the 3 pieces of filter paper in ~~the~~ ^ Petri dish. ~~Next, 1, .1, and .01 concentrations of GA were made using a serial dilution of 10 μM GA and 10 mM HEPES-EGTA-Ca2+.~~ Then 3.5 mL of ~~buffer~~ ^ or ^ hormone ~~were~~ ^ added to each of the Petri dishes according to the following table.

not a recipe

^with
^were placed
^side
^a

^10 mM HEPES-EGTA-Ca^{2+} buffer / 1, 0.1, and 0.01 μM GA / solution was

Plate	Seeds	Treatment
A	Whole seeds	3.5 mL buffer
B	Half seeds	3.5 mL buffer
C	Half seeds	3.5 mL .01 μM GA
D	Half seeds	3.5 mL .1 μM GA
E	Half seeds	3.5 mL 1 μM GA
F	Half seeds	3.5 mL 10 μM GA

num format (fix every time)

	The Petri dishes were wrapped in foil and placed in
symbol: ^15 °C	a ~~15 degrees Celsius~~ ^ chamber for one week. After 1 week ~~the plates were collected and~~
1 g singular: ^was	approximately 1 g of seeds ~~were~~ ^ put into a mortar ~~that was~~ on ice. The seeds were pulverized with 10 mL cold 10 mM citric acid-sodium citrate buffer at pH 5 for 5 min. The pulverized seeds were then placed in a centrifuge tube and the mortar was rinsed with 15 mL ice cold buffer to get the seed product into the
wordy	tube. ~~This was done separately for each one of the plates.~~ The six samples were centrifuged for 10 min at
abbr: ^2 °C	15,000G at ~~2 degrees Celsius~~ ^. ~~The supernatant was collected.~~ One mL of each ~~one of the six~~ supernatant was ~~added to six tubes at room temperature. Then~~
^mixed with	~~in 20 second intervals~~ ^ 2 mL of .025% starch ~~was added to each tube successively~~. After 120 sec, 7 mL of 1 N HCl was added to ~~the first tube and then~~
^each sample, followed by 1	~~down the line every 20 seconds. One~~ ^ mL of Lugol's iodine solution ~~was added to each tube~~. Lastly, the
wc: ^sample	absorbance ~~value~~ of each ~~tube~~ ^ was taken with a Spec 20 spectrophotometer at 580 nm. The ~~experimental results of~~ starch ~~concentrations were compared to a standard curve.~~ ~~Using a 0.05%~~
^standards were prepared by combining 2 mL of / starch solution with	~~stock solution, a serial dilution was used to make~~ ^ .025%, .0125%, .006% and .003% ^ ~~concentrations. Two mL of each solution was poured into 5 different tubes. A sixth tube (blank) was filled with 2 mL water. Then~~
sp	1 mL buffer, 7 mL HCl, annd 1 mL Lugol's iodine solution ~~was added to each tube~~. The absorbance
wordy	~~value for each tube~~ was determined ~~by a Spec 20 spectrophotometer~~ at 580 nm. ~~Absorbance was plotted on the y-axis and percent starch was plotted on the~~
don't preview	~~x-axis. A best fit trend line was inserted to determine the equation of the standard curve.~~
	Results
interp	~~The results from the experiment support the hypothesis that the digestive enzyme α-amylase production is dependent on the presence of gibberellic acid (GA) in germinating seeds. α-amylase promotes plant growth growth by digesting starch in the seed's endosperm. The higher the gibberellic acid content,~~

~~the higher the amount of α-amylase production and starch digestion.~~

~~The experiment included six groups of germinating seeds. Whole seeds in a 3.5 mL buffered solution served as the positive control because their embryos naturally produce gibberellic acid. They were expected to produce α-amylase and thus exhibit starch digestion. The negative control seeds were the endosperm halves of seeds in a 3.5 mL buffered solution because they had no way of synthesizing or obtaining gibberellic acid. The negative control group was not expected to produce any α-amylase or digest as much starch as the other groups. The other four groups of endosperms received treatments of 3.5 mL GA in varying concentrations.~~

~~The starch standard curve indicates a positive relationship between light absorbance at 580 nm and starch concentration.~~ The equation of the standard curve (Figure 1) was used to calculate the percentage of starch remaining in the endosperm of the seeds after each treatment. The absorbance values determined for each of the six treatments were substituted into the equation for y to determine the remaining starch concentration (x).

unnec: absorbance is supposed to be proportional to concentration by Beer's Law

The whole seed positive control and the GA-treated half seeds all exhibited lower percentages of ~~absorbance~~ ^ than the negative control. The ~~absorbance level and~~ starch (%) decreased as the concentration of the 3.5mL μM gibberellic acid (GA) increased ~~per treatment. The whole seeds had 0.0216% of their starch remaining, while the 0.01 μM GA, 0.1 μM GA, 1 μM GA, and 10 μM GA had 0.0235%, 0.0207%, 0.0096%, and 0.0079% starch remaining, respectively. The negative control had no GA exposure, and had a higher (0.0247%) percentage of its starch remaining.~~

wc: ^starch

too much detail

The starch remaining per gram of seed was calculated to eliminate any error as a result of the varying weights of the sample. To determine the amount of starch digested per gram of seeds for each treatment, the percentage of starch digested in each sample was divided by its weight in grams (Figure 2).

move this ¶ ahead of previous one

fig format

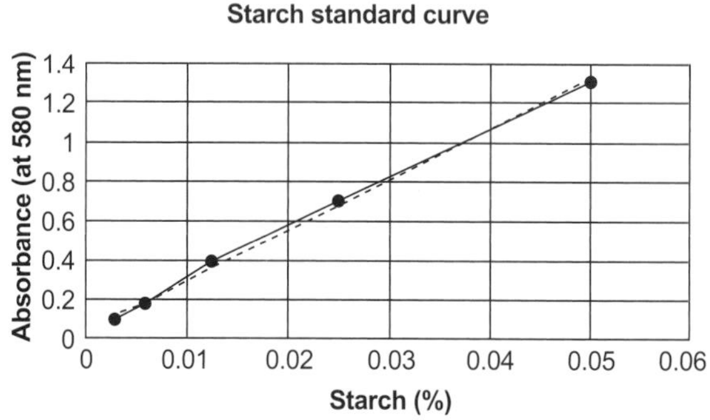

unnec: can read axis labels

Figure 1 Starch standard curve ~~describing the relationship between starch remaining and light absorbance~~

past tense

interp belongs in Discussion section

As figure 2 shows, the starch remaining per gram of seeds is lower in the samples with higher concentrations of GA. The hypothesis that gibberellic acid promotes α-amylase production and starch digestion is supported by the fact that the treatments with higher concentrations of gibberellic acid consumed more starch than even the positive control group.

unnecessary to rehash methods

~~The α-amylase activity measured the rate at which α-amylase digested starch in the endosperm. Each sample was mixed with 25 mL 10mM citric acid-sodium citrate buffer of pH 5.0 and then pulverized into a suspension. The suspensions were centrifuged~~

fig format

legend needs to identify the treatment groups

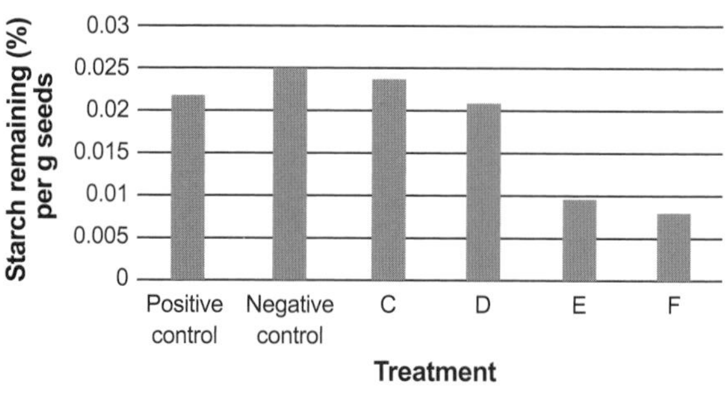

to make title self-explanatory: ^GA ^barley

Figure 2 Effect of ^ treatments on the percentage of starch remaining per gram of ^ seeds

~~to derive supernatants containing the enzymes created in the reaction with citric acid-sodium citrate buffer. The supernatants of each sample were treated with 1 mL of 0.025% starch solution each and allowed to react for 120 seconds. The alpha amylase enzymes in the suspensions digested the starch solutions. The α-amylase activity was calculated by dividing the starch digested per gram of seeds by the reaction time (Figure 3).~~ γ

~~The data in~~ Figure 3 ~~show that the amount of starch digested was dependent on the amount of α-amylase present in the supernatant sample.~~ Treatments with higher concentrations of gibberellic acid ~~appeared to have~~ ^ the most α-amylase activity ^. The negative control had the least amount of activity at 3.16x10-6 starch digested/g/seeds/sec. This greatly contrasts with sample F, with 3.5 mL 10 μM GA and 1.43×10^{-4} starch digested/g/seeds/sec. The positive control also had a relatively low amount of α-amylase activity compared to the high concentration treatments. ~~The trend in Figure 3 shows that the α-amylase activity increased proportionately with the concentration of GA in each sample. The results shown in Figure 3 support the notion that GA content indirectly affects starch digestion.~~

meaning γ

past tense: ^had

^( )

super exp

rep

meaning

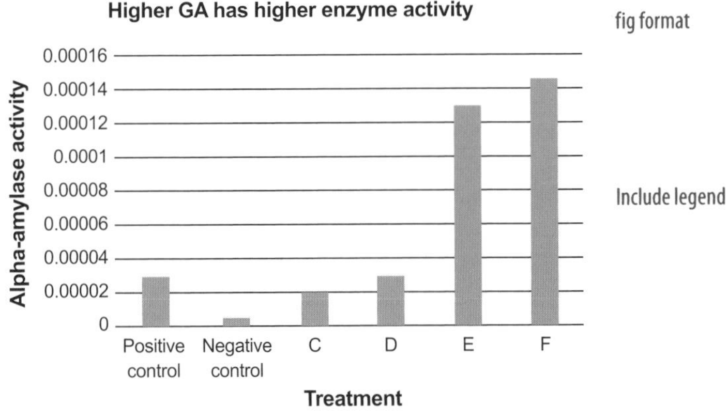

fig format

Include legend

Figure 3 Effect of ^ treatments on α-amylase activity in ^ seeds

^GA

^barley

wordy

num format (use sci notat)

move this ¶ after next one. Disc structure specific to broad.

cit format

Discussion

The results shown in Figure 2 support the idea that amount of exogenous GA affects starch digestion in grass seeds through its induction of α-amylase production. As seen in Figure 2, higher concentrations of GA allowed seeds to digest higher percentages of starch. Samples E and F, with higher GA concentrations of 1 μM and 10μM, digested the highest percentages of starch at .015495% and .0171%, respectively.

The experiment involved supplying endosperms with varying amounts of GA in order to simulate the embryo's natural production of α-amylase. The α-amylase enzyme is created in the aleurone layer of the seed and digests the available starches in the endosperm. The starches are broken down into glucose and nourish the seed in germination. Without gibberellic acid, the seed would not be able to process the available starches for nourishment.

The concentration of exogenous GA directly affected the α-amylase activity of each sample as well. Figure 3 shows that α-amylase activity was higher in samples with higher concentrations of exogenous GA. α-amylase activity increased according to the increase in GA. Sample C, with a 3.5 mL .01 μM GA concentration, yielded lower levels of α-amylase activity than even the positive control. Sample C's low α-amylase activity due to highly diluted GA provides evidence supporting the hypothesis that α-amylase activity depends on the presence of gibberellic acid.

Our results are similar to those reported by Skadsen for two cultivars of barley, Morex and Steptoe. He set up plates with embryo-less half seeds treated with 0, .01 μM and 1 μM GA. After 5 days he measured α-amylase activity. His results were similar to ours for the Morex cultivar in that higher GA concentrations resulted in higher α-amylase activity. This trend was not evident in the Steptoe cultivar, possibly due to an inhibitor in the endosperm (Skadsen).

Production of α-amylase and starch degradation in the seed may have been affected by several sources of error. The separation of endosperm from embryo was performed without careful measurements; any embryo left attached to its endosperm may provide the seed with endogenous GA not accounted for in the calculations. In the negative control group, the presence of embryonic remains may have contributed to a slight amount of α-amylase activity (3.16x10-6 starch digested/g/seeds/sec) observed in Figure 3. Another factor contributing to error includes timing of each reaction with starch. The reaction began when 2 mL of .025% starch was added to each sample. Each reaction was intended to endure 120 seconds until stopped by the addition of 1 N HCl. If the 1 N HCl was not added at the correct time, the amount of actual starch digested would be inconsistent with the calculated value.

good error analysis

super exp

References

Koning, Ross E. 1994. Seeds and Seed Germination. Plant Physiology Information Website. http://plantphys.info/plants_human/seedgerm.shtml. (accessed 11-20-2016).

Paleg L.G. Physiological effects of gibberellic acid. II. On starch hydrolyzing enzymes of barley endosperm. Plant Physiology 1960; 35(6): 902-906.

Skadsen RW. ^ Aleurones from a Barley with Low α-Amylase Activity Become Highly Responsive to Gibberellin When Detached from the Starchy Endosperm. Plant Physiol. ~~(1993)~~ 102: 195-203

Tanaka Y, Akazawa T. Plant Physiology "Alpha-amylase isozymes in gibberellic acid-treated barley half-seeds" 1970 ~~Apr~~; 46(4): 586^

cit missing

N-Y ref format: year follows author's name ^1993

Write out journal title

ref format: year follows authors' names

^-591; inclusive pages

Go to the **COMPANION WEBSITE** • **sites.sinauer.com/knisely5e**
for samples, template files, and tutorial videos

Chapter 7

Poster Presentations

Posters are a means of communicating preliminary research results both orally and visually. **Poster sessions** are often held in large exhibit halls at national meetings, where presenters stand next to their poster and discuss their work with interested attendees. The informal conversations provide great networking opportunities for young scientists.

Why Posters?

Scientists who attend poster sessions constitute a much larger audience than the one attracted to a journal article on a particular topic. Thus, your goal is to produce a poster that not only attracts experts in your subdiscipline, but also the much larger group of scientists with tangential research interests. The latter group provides a unique opportunity for you to learn about applications of your work to other research areas (and vice versa), spurs scientific creativity, and prompts you to apply an interdisciplinary approach to problem-solving.

Posters are *not* papers; they rely more on visuals than on text to present the message. It is not necessary to supply as many supporting details as you would for a paper, because you (the author) will be present to discuss details one-on-one with interested individuals. Too much material may even discourage individuals from reading your poster.

An appropriate poster presentation should fulfill two objectives. First, it must be esthetically pleasing to attract viewers in the first place. Second, it must communicate the objectives, methods, results, and conclusions clearly and concisely.

Poster Format

Posters can be constructed on poster boards or created in PowerPoint and printed on large-format paper. If you are presenting at a professional society meeting, check with the conference organizer regarding size and other requirements. For poster sessions in your class, ask your instructor about appropriate materials and sizes.

Layout

Make a rough sketch of the layout before you begin. Position the title and authors prominently at the top of the poster. Then arrange the body of the poster in 2, 3, or 4 columns, depending on whether the orientation of the poster is portrait or landscape. If you don't want to make your poster from scratch, check if your institution provides poster templates. Scientific poster templates are also available on the Internet (see, for example, Alley [2016], Purrington [2016], and other sources listed in the Bibliography).

Appearance

The success of a poster presentation depends on its ability to attract people from across the room. Interesting graphics and a pleasant color scheme are good attention-getters, but avoid "cute" gimmicks. Present your poster in a serious and professional manner so people will take your conclusions seriously.

Organize the sections so that information flows from top left to bottom right. Align text on the left rather than centering it. The smooth left edge provides the reader with a strong visual guide through the material.

Avoid crowding. Large blocks of text turn off viewers; instead, use bullets to present your objectives and conclusions clearly and concisely. Use blank space to separate sections and to organize your poster for optimal flow from one section to the next.

Make sure your text is legible. Use black font on a white or light-colored background for best contrast.

Use appropriate graphics that communicate your data clearly. Use three-dimensional graphs *only* for three-dimensional data. Make sure that the numbers and labels on figures are large and legible and free of typos.

Colored borders around graphics and text enhance contrast, but keep framing to a minimum. **Framing** is the technique whereby the printed material is mounted on a piece of colored paper, which is mounted on a piece of different-colored paper to produce colorful borders. Use borders judiciously so that they do not distract from the poster content.

When using poster boards, use adhesive spray or a glue stick to affix text and figures. These tend to have fewer globs and bulges than liquid glues.

Font (type size and appearance)

Keep in mind that most visitors will be reading your poster from 3 to 6 feet away. Sans serif fonts like **Arial** are good for titles and labels, but serif fonts like **Times** and **Palatino** are much easier to read in extended blocks of text. The **serifs** (small strokes that embellish the character at the top and bottom) create a strong horizontal emphasis, which helps the eye scan lines of text more easily.

Format the title of the poster in sentence case or title case in 72 point or larger **bold**. In **sentence case**, only the first letter of the first word is capitalized. In **title case**, the first letter of the most important words is capitalized. These styles are easier to read and take up less space than all caps. Keep the case of case-sensitive words, such as pH, cDNA, or mRNA, as is. Limit title lines to 65 characters or less.

Times 72

Arial 72

Authors' names and affiliations should be at least 48 point **bold**, serif font, title case:

Times 48 Pt

The section headings can be 32-48 point **bold**, serif font, sentence case or title case:

Times 28 Pt

The text itself should be no smaller than 24 point, serif font, sentence case, and *not* in bold:

Times 24 pt

In terms of readability, it's better for the font to be larger rather than smaller.

Nuts and bolts

Ask the conference organizer (or your instructor) about how posters will be displayed at the poster session. Some possibilities include a pinch clamp on a pole, an easel, a table for self-standing posters, and cork bulletin boards to which posters are affixed with pushpins.

Making a Poster in Microsoft PowerPoint

Many scientists use PowerPoint to prepare posters for professional society meetings because this software is readily available; the font, color scheme, and layout can be customized by the authors; the content can be revised easily; the electronic file format facilitates collaboration among colleagues in different locations; and the final product looks professionally done. Because you are at the mercy of your computer when making posters in PowerPoint, however, save your file early and often!

The entire poster is made on one PowerPoint slide, whose size is set in the **Slide Size** dialog box. Text is written in text boxes, which are inserted along with images and graphs on the slide as desired. The following sections describe some of the basic tasks you will carry out when making your poster in PowerPoint 2013.

Design

1. Open a blank PowerPoint presentation.
2. Click **Design | Slide Size**. Windows users click **Custom Slide Size**; Mac users click **Page Setup**. Then enter the poster dimensions (Figure 7.1). Stay within the dimensions specified by the conference organizer, and check with whomever is printing your poster about any size restrictions. Common sizes include 30 × 40″, 36 × 48″, and 56 × 41″. Portrait or landscape orientation will be selected automatically based on how you enter the numbers. An advantage of landscape is that visitors don't have to look up or down as far to view the whole poster.
3. Use black text on a white or light-colored background for best contrast.

Adding text, images, and graphs

- To add text to your poster, click **Insert | Text Box** for each section or block of text. Aim for a consistent look by using the same family of fonts. Adjust the font size for the title, authors' names

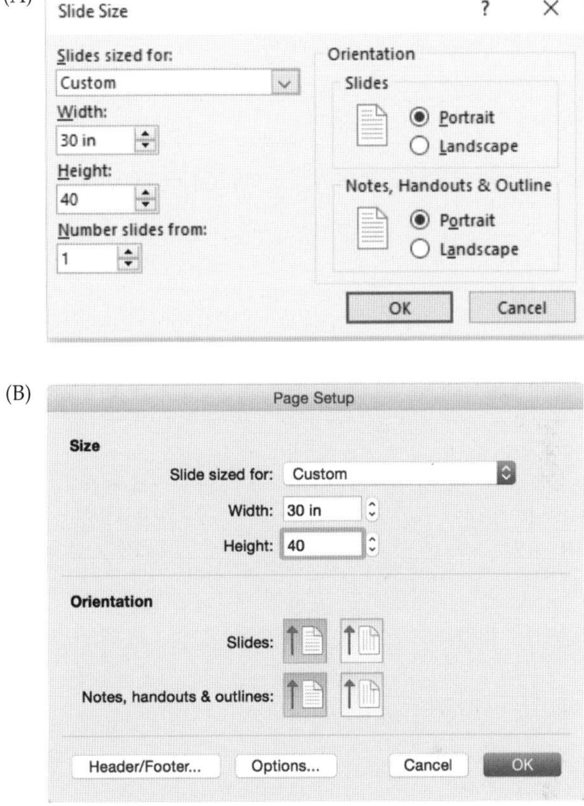

Figure 7.1 Poster dimensions are specified in the **Slide Size** dialog box on the **Design** tab. (A) PowerPoint 2013. (B) PowerPoint for Mac 2016.

and affiliations, section headings, and text by making the appropriate selections on **Home | Font**. To change the properties of the text box itself, Windows users click the **Drawing Tools Format** tab on the Ribbon. Mac users click **Shape Format**.

- To insert pictures, click **Insert | Pictures**.
- To insert a graph, copy it from Excel and paste it into your PowerPoint slide as a picture. When you drag on one of the corners, the font size of the axis labels automatically increases with the size of the graph. With regular paste, you would have to increase the font size of each block of text manually.

- To make your own graphics, see "Working with shapes" in Appendix 3.
- Use the **Zoom slider** in the lower right corner to enlarge the section of the poster you're working on.

Aligning objects

1. To center the title and authors on the slide, hold down the **Shift** key, click each text box, and click **Align | Align Center** on the **Drawing Tools Format** tab (Windows) or **Shape Format** tab (Mac) (Figure 7.2).
2. Now display the ruler by clicking **View | Ruler**. Drag or nudge the grouped text boxes so that the center handle lines up with the zero on the horizontal ruler. To nudge an object, click it and hold down the **Ctrl** (Windows), or **Command** (Mac) key while pressing the arrow keys.
3. To align the left edges of text boxes and images, hold down the **Shift** key, click each object, and then click **Align Left** on the **Drawing Tools Format** tab (Windows) or **Shape Format** tab (Mac).

Proofread your work

Print a rough draft of your poster on an 8.5×11" sheet of paper. To do so, click **File | Print**. Windows users, under **Settings**, click the down arrow

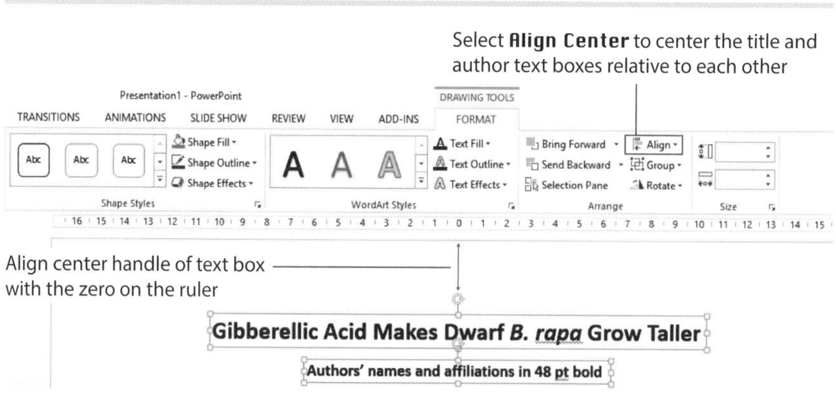

Figure 7.2 Center the title and authors' names and affiliations by first clicking **Align Center** and then aligning the center handle of the text box with the zero on the horizontal ruler. This screenshot is from PowerPoint 2013. The menus for PowerPoint for Mac 2016 are similar.

next to **Full Page Slides** and check that **Scale to Fit Paper** is selected (this is the default for both Windows and Mac). Check for typos and grammatical errors and evaluate the layout and overall appearance of your poster. If the print is not large enough to read on the draft, it will probably be hard to read from a distance on the full-sized poster as well. Adjust the font size under **Home | Font**.

Final printing

Check with your local print shop concerning the electronic format of the poster file. PowerPoint (.ppt or .pptx) may be acceptable, but Portable Document Format files (PDFs) are becoming the printing industry standard. PDFs tend to be smaller than the source file, and all of the fonts, images, and formatting are retained when the document is printed.

Poster Content

Posters presented at professional society meetings should be organized so that readers can stand 10 feet away from the poster and get the take-home message in 30 seconds or less. Because of the large number of sessions (lectures) and an even larger number of posters, conference participants often experience "sensory overload." Thus, if you want your poster to stand out, make the section headings descriptive, the content brief and to the point, and the conclusions assertive and clear.

Posters for a student audience in the context of an in-house presentation should follow the same principle of brevity, but may retain the sections traditionally found in scientific papers. These include:

- Title banner
- Abstract (optional—if present, it is a summary of the work presented in the poster)
- Introduction
- Materials and Methods
- Results
- Discussion or Conclusions
- References (less comprehensive than in a paper)
- Acknowledgments

Title banner

Use a short, yet descriptive, title. This is the first and most important section for attracting viewers, so try to incorporate your most important conclusion in the title. For example, **Gibberellic Acid Makes Dwarf *B. rapa***

Grow Taller is more effective than **Effect of Gibberellic Acid on Dwarf *B. rapa*.**

The title banner should be at the top of the poster and in 72 point (or larger) bold font, sentence case or title case, 65 characters or less on a line. Underneath the title, include the authors' names and the institutional affiliation(s). Use at least 48 point bold, title case for the authors' names.

Introduction

Instead of using the conventional "Introduction" heading found in papers, use a short, attention-grabbing question such as "What makes dwarf plants short?" Then provide the minimum amount of background information on the topic and state your hypothesis. A bulleted list may be a good way to present some of this information.

Materials and methods

Summarize the methods in enough detail so that someone with basic training could repeat your experiment. If possible, use pictures to show the experimental setup and flow diagrams or a bulleted list for the procedure.

Results

The Results section of a poster consists mostly of visuals (pictures and graphs). Announce each important result with a heading *above* the figure. It is not necessary to write a detailed caption below the figure, as you would in a scientific paper. Use a bulleted list to point out general patterns, trends, and differences in the results. Poster viewers do not have the time to read the results leisurely, as they would with a paper.

Figures may contain some statistical information including means, standard error, and minimum and maximum values, where appropriate. Make the data points prominent and use a simple vertical line without crossbars for the error bars. Make the font size on the axis numbers and axis titles at least 24 pt. When there are multiple data sets on a graph, instead of using a legend, place a descriptive label next to each line. This approach allows visitors to identify the conditions more quickly. Avoid using tables with large amounts of data.

Images are a great addition to a poster, but only if they are sharp. Picture quality is determined by the size (number of pixels) of the digital image and the printer resolution. Various sources recommend a printer resolution of at least 240 pixels per inch (ppi) to print quality photos. That means for a

5″ × 7″ print, the image would have to be at least 1200 × 1680 pixels to attain the desired resolution on paper. To check the size of an image, right-click it and select **Properties**. Be careful when resizing pictures. Enlarging an image that does not have enough pixels will result in a blurry picture.

Edit the text ruthlessly to remove nonessential information about the visuals. A sentence like "The effect of gibberellic acid on *B. rapa* is shown in the following figure" is nothing but deadwood. On the other hand, "*B. rapa* plants treated with gibberellic acid grew taller than those receiving only water" informs the reader of the result.

Remember to leave blank space on a poster. Space can be used to separate sections and gives the eyes a rest.

Discussion or Conclusions

Most visitors will be especially interested in the Conclusions section, so make sure this section is featured prominently on your poster. This is where you discuss your results. Use bullets to state the take-home message and provide the experimental evidence. State whether or not your results support your hypothesis. Were there any unexpected or surprising results? What questions remain unanswered? If you were to repeat the study, what would you do differently?

Literature citations

In posters as in scientific papers, it is imperative to acknowledge your sources, which are cited most often in the Introduction and Discussion/Conclusions sections. Because posters are informal presentations and the author is present to provide supporting details, the References section of a poster can be less comprehensive than that in a paper. Some ways to save space while still acknowledging your sources are to use the citation-sequence system and to abbreviate the end references, for example, by eliminating the article title or by using DOIs and URLs for online publications. However, for in-class poster presentations, your instructor may ask you to hand in a correctly formatted reference list or bibliography. For in-text and end reference format, see the section "Documenting Sources" in Chapter 4.

Acknowledgments

In this section, the author acknowledges organization(s) that provided funding and thanks technicians, colleagues, and others who have made significant contributions to the work.

Presenting Your Poster

Prepare a 3–5 minute talk explaining your poster. This "elevator pitch" should be an engaging summary of why and how you did the work, your most important findings, and your conclusions. Even if you don't actually give your pitch at the poster session, verbalizing the content of your poster is good preparation for the one-on-one conversations you will have with conference participants. Bring along a stack of business cards or handouts with your contact information to give to interested visitors for possible future collaboration.

Evaluation Form for Poster Presentations

Good posters are the product of creativity, hard work, and feedback at various stages of the poster preparation process. When you present a poster to your class or at a professional society meeting, participants may be asked to evaluate your poster using a form such as the "Evaluation Form for Poster Presentations" available at <http://sites.sinauer.com/Knisely5E>.

When you are in the position of evaluator, make the kinds of comments you would find helpful if you were the presenter. As you know, feedback is most likely to be appreciated when it is constructive, specific, and done in an atmosphere of mutual respect.

Sample Posters

Examples of posters along with reviewer comments are posted on <http://sites.sinauer.com/Knisely5E>. What do you like (or not like) about each poster? Use the evaluation form to determine how well the authors have met the requirements of good poster design.

Go to the **COMPANION WEBSITE** • **sites.sinauer.com/knisely5e**
for samples, template files, and tutorial videos

Chapter 8

Oral Presentations

Scientific findings are communicated through journal articles, poster presentations at meetings, and oral presentations. Oral presentations are different from journal articles and posters because the speaker's delivery plays a critical role in the success of the communication.

As a student, you have been on the receiving end of oral presentations for a number of years and probably have a pretty good idea of what makes a talk memorable. For example, presentations in which the speaker appears to be unprepared, uncomfortable speaking in public, and unfamiliar with the level of his or her audience are quickly forgotten regardless of the content. On the other hand, successful presentations tend to have the following characteristics:

- Speakers who show enthusiasm for their topic
- Speakers who establish a good rapport with the audience
- Information that is well organized, leaving listeners satisfied that they understand more than they did before
- Visuals that are simple and legible, and help listeners focus on the important points without drawing attention away from the speaker

Structure

Oral presentations are not unlike scientific papers in their structure, but they are much more selective in their content. As in a scientific paper, the **introduction** captures the audience's attention, provides background information on the subject, and identifies the objectives of the work (Table 8.1). The **body** is a condensed version of the Materials and Methods and Results sections, and contains only enough detail to support the speaker's conclusions. If the focus of the talk is on the results, then the speaker spends less time on the methods and more time on visuals that highlight the findings. The visuals should be simpler than those prepared for a journal article, because the audience may only have a minute or two to digest

TABLE 8.1 Comparison of the structure of an oral presentation and a journal article

Oral Presentation	Journal Article
	Abstract
Introduction	Introduction
Body	Materials and Methods, Results
Closing	Discussion
	References
	Acknowledgments

the material in the visual (in contrast to a journal article, where the reader can re-read the paper as often as desired). Finally, the **closing** is comparable to the Discussion section, because here the speaker summarizes the objectives and results, states conclusions, and emphasizes the take-home message for the audience. The closing may include an acknowledgments slide on which the speaker lists sources of funding and recognizes research advisers, collaborators, and others who helped with the work.

There is no distinct abstract or References section in an oral presentation. If the presentation is part of a meeting, an abstract may be provided in the program. References may be given in an abbreviated form at the bottom of a slide or below visuals. If the **slide deck** (the slides with any notes) will be shared, full references can be included in the notes.

Plan Ahead

Before you start writing the content of your presentation, make sure you know the following:

- **Your audience.** What do they know? What do they want to know? If possible, ask some carefully chosen questions to assess their experience and motivation.
- **Why you are giving the talk**, not just why you did the experiment. Talks may be **instructional**—the speaker wants listeners to leave with new knowledge and skills. Talks may be **informational**—the speaker presents research to fellow scientists at a seminar or journal club meeting. Or talks may be **persuasive**—the speaker may try to get listeners to provide funding or to adopt a certain point of view on a controversial topic.
- **Your speaking environment.** How much time do you have? How large is the audience? What presentation equipment is available

(chalkboard, overhead projector, computer with projection equipment, Internet connection)?

- **How the slide deck will be used** beyond the presentation. If you intend to share the deck, include explanatory text that would enable your virtual audience to understand the slides without you, the speaker, being present.

Prepare the First Draft

An oral presentation will actually take you longer to prepare than a paper. D'Arcy (1998) breaks down the steps as follows:

- Procrastination 25%
- Research 30%
- Writing and creating visuals 40%
- Rehearsal 5%

To overcome procrastination, follow the same strategy as for writing a laboratory report: write the body first (Materials and Methods and Results sections), then the introduction, and finally the closing.

- Write the main points of each section in outline form.
- Use science news articles, review articles, and journal articles to provide background in the introduction and supporting references in the closing. Use sources that are appropriate and engaging for your audience.
- Determine how much detail to present on each topic. Consider the level of your audience and how much time you have for the presentation.
- Transfer each topic on your outline to its own slide in your presentation software program. There are several programs on the market, but Microsoft PowerPoint is the most popular. PowerPoint has been widely criticized for its bullet-laden templates, but it is really up to you, the presenter, to choose a suitable layout and design for your slides.
- Do a preliminary, timed run through your presentation. Eliminate nonessential information to stay within the time limit.

Make the Slides Audience-Friendly

The first draft of your presentation is all about you, the speaker. The topic-bullet approach is a convenient way for you to organize your talk, but text-heavy slides are a sure way to turn off your audience. Audience members simply cannot read line after line of incomplete thoughts and simultaneously listen to what you are saying. Therefore, it is imperative that you make your slides audience-friendly. Studies have shown that a combination of *hearing* the speaker, *reading* a simple, complete statement, and *seeing* a visual increases audience comprehension and retention (Alley 2016; Nathans-Kelly and Nicometo 2014; McConnell 2011).

Here are some suggestions for making your slides audience-friendly. Of course, if you were given specific instructions for your presentation, then those take precedence over these suggestions.

Focus on one idea at a time

Evaluate each slide critically. Ask yourself, "What is the point of this slide?" If you find yourself listing several key points, consider splitting up the information over several slides. **One idea per slide** is easier for listeners to comprehend, because they can focus all their attention on just one topic, instead of having to figure out how multiple topics are related.

Write complete thoughts

Replace ambiguous slide titles like "Introduction" with complete thoughts that eliminate guesswork. Give audience members the take-home mes-

(A) Script for the speaker

Introduction

- The femur is an unpaired bone that has to bear the entire weight of an animal during running.
- Load is proportional to mass.
- Cross-sectional area has to bear the load.
- We wanted to see how the cross-sectional area changes with the mass of the animal.

(B) Audience-centered slide

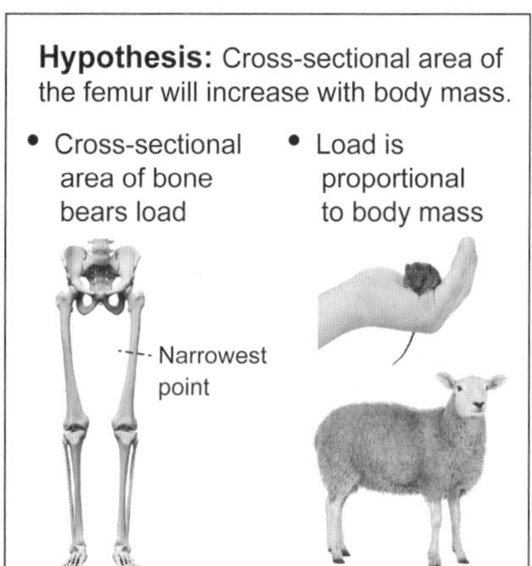

Hypothesis: Cross-sectional area of the femur will increase with body mass.
• Cross-sectional area of bone bears load
• Load is proportional to body mass
Narrowest point

Figure 8.1 Speaker-friendly (A) and audience-friendly (B) ways to construct a slide. The large amount of text on Slide A will have the audience reading instead of listening to the speaker. The simple, complete statement on Slide B tells the audience what is important. Well-chosen pictures, displayed sequentially for emphasis, support the statement. Photos © leonello, palenka19, and GlobalP/iStock.com.

sage; don't make them waste valuable time wondering what they are supposed to remember (Figure 8.1). The **complete-sentence style titles** should be written in active voice and sentence case and have consistent punctuation and alignment. Titles should be no longer than two lines when written in 32-40 point font.

Use more visuals and fewer words

As much as possible, **replace text with a visual** that supports the title sentence (Figure 8.2). It takes less time and effort to comprehend a picture than to read and process the words that the picture illustrates. Be careful not to overwhelm the audience with details. Present the minimum amount of information that listeners need to understand the topic.

Keep graphs simple

Make the numbers and the lettering on the axis labels large and legible. Choose symbols that are easy to distinguish and use them consistently in all of your figures. For example, if you use a circle to represent "Wild type in glycerol" in the first graph, use the same symbol for the same condition

(A) Script for the speaker

Methods

- Measured femur length in mm with a ruler
- Measured diameter of femur in mm with calipers

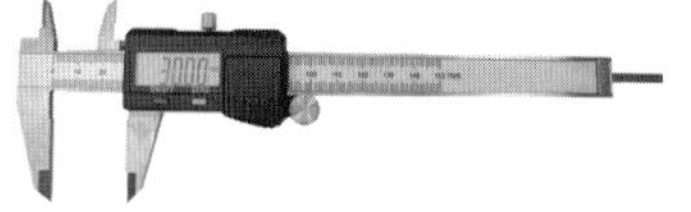

- Calculated cross-sectional area (Diameter/2)^2*(pi)

(B) Audience-centered slide

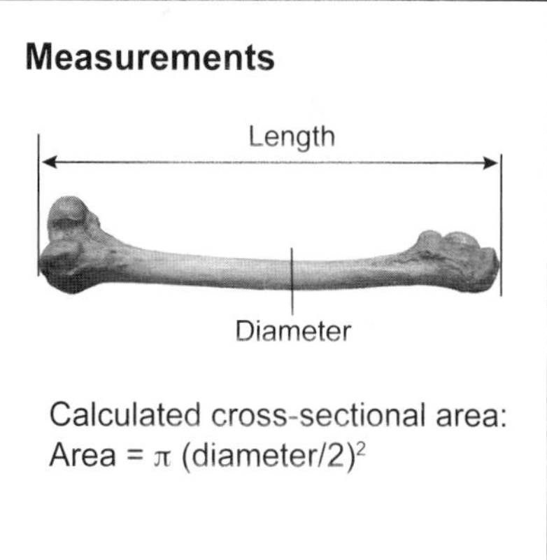

Figure 8.2 Speaker-friendly (A) and audience-friendly (B) ways to construct a slide. The audience can look at a simple, labeled picture and understand what was measured much faster than reading a description of the procedure. Irrelevant details, such as the instrument used to make the measurements, are omitted from the slide. Photos © lfness and Uber-Anatomy/iStock.com.

in the rest of your graphs. Instead of following the conventions used in journal articles, modify the figure format as follows (Figure 8.3):

- Position the title above the figure and don't number it. The title should reflect your most important finding.

(A)

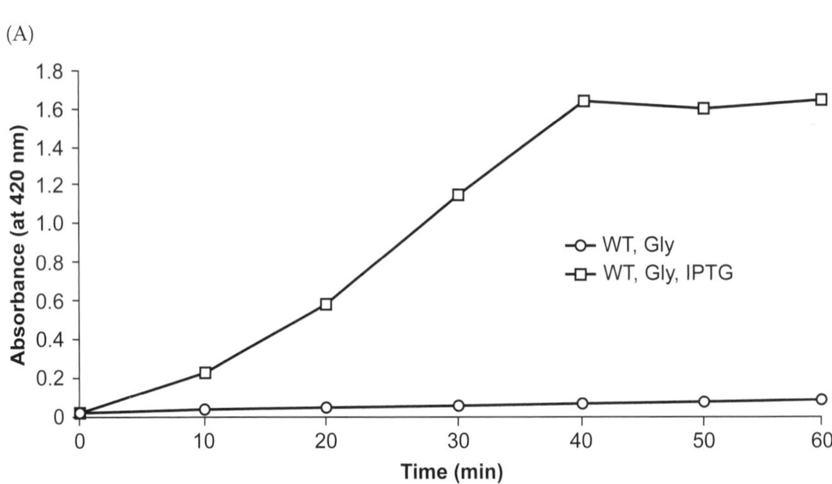

Figure 1 Beta-galactosidase production in wild type *E. coli* grown in glycerol media. Inducer (IPTG) was added to the culture at the start of the assay.

(B)

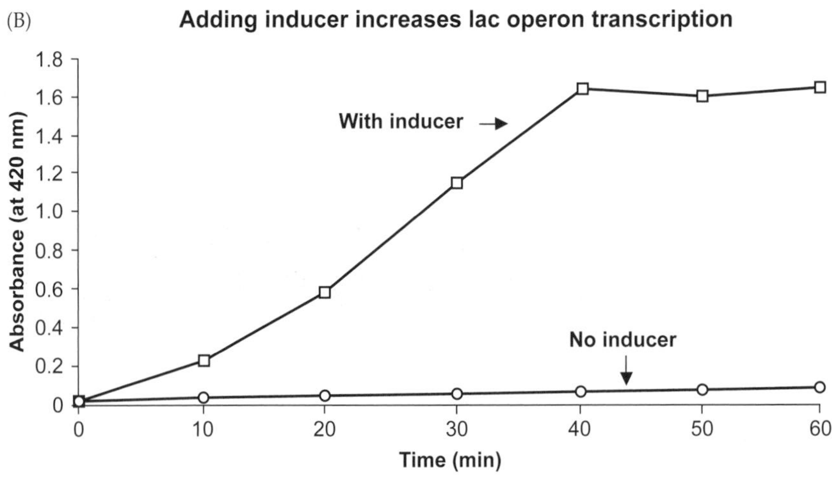

Figure 8.3 Example of figure formatting for (A) a journal article and (B) an oral presentation.

- Label each line instead of identifying the data symbols with a legend. The labels may be made using text boxes in Excel, Word, or PowerPoint.

Both of these modifications make it easier for the listener to digest the information within the short time the graph is displayed during the presentation.

Make text easy to read

Use fonts and a color scheme to make text easy to read from anywhere in the room.

Use at least 24 pt font for bullet points, labels, and axis titles

Use sans serif font for slides for easier reading.

Use serif font for handouts.

Use color sparingly in text slides: Nearly 10% of the male population is color blind.

A bright color, **boldface**, or a larger font, when used with discretion, are good ways to emphasize a point.

Use black type on white background for best contrast. The light background also helps keep the audience awake and allows you to see your listeners' faces.

In a large room, light-colored type on a dark background makes characters look bigger.

Provide ample white space

Use white space liberally, especially around the edges of the slide. If the slides are formatted for wide screen (16:9) and the projection equipment is set for standard format (4:3), content near the edges may be cut off.

Use animation feature to build slide content

Sometimes it may be necessary or desirable to present a complicated concept, process, or result on a single slide. However, rather than displaying all of the components at once, the speaker "builds" content by revealing only one element at a time. This technique is often applied to overview and conclusion slides, on which bulleted items appear one after the other, allowing the audience to focus on one point at a time while the speaker is talking about it. This effect is produced with PowerPoint's animation feature. For instructions on how to apply animations to your PowerPoint presentations, see pp. 262–263 in Appendix 3.

Shapes can also be animated to add emphasis. For example, the speaker may call attention to a certain point on a graph with an arrow or highlight important information with a red circle or box. Using animations eliminates any ambiguity caused by a pointer being waved in the general direction of the visual.

Appeal to your listeners' multiple senses

Use images from the Internet (with appropriate acknowledgment), recordings of heart sounds or bird songs, and videos if they help make your point. Make sure these "extras" do not detract from your take-home message and that they fit within the time limit of your presentation.

Deal proactively with lapses in audience attention

The attention level of an audience is highest at the beginning of a presentation, decreases in the middle, and then recovers at the end. In a typical scientific presentation, that means the audience will be the most distracted during the most technical part of the talk.

Experienced presenters are aware of this problem and deal proactively with lapses in audience attention. For example, professors may take a break partway through their lecture and have students do an activity that involves small-group discussion, or a reflective writing exercise, or a clicker quiz to assess comprehension. Forcing students to take an active role in the class has the dual benefit of clearing up misconceptions and allowing those who zoned out to catch up. In research talks, you may have noticed that after speakers present an especially complex topic, they pause and review the most important points. These approaches for recapturing audience attention are analogous to a diver (the speaker) swimming to deep waters and then returning to the surface after a while, as illustrated in Figure 8.4. The idea is that the speaker dives into increasingly difficult material, but periodically brings the audience up for air. During this recovery phase, the audience gets another chance to hear the key points,

Figure 8.4 "Data dives" approach for keeping the audience on track during the most technical portion of an oral presentation. The speaker-diver takes audience members into depth on a topic, but then periodically brings them up for air by recapping the most important ideas. Adapted from McConnell 2011. Diver © majivecka/iStock.com.

and the respite helps prevent those whose attention has lapsed from becoming hopelessly lost.

Rehearsal

After you have prepared your presentation, you must practice your delivery. Give yourself plenty of time so that you can run through your presentation several times and, if possible, do a practice presentation in the same room where you will hold the actual presentation. Here are some other tips:

- Go to a place where you can be alone and undisturbed. Read your presentation aloud, paying attention to the organization and especially the connecting sentences. Does one topic flow into the next, or are there awkward transitions? Revise both the PowerPoint and your speaker notes as needed. Apply the revision strategies described in Chapter 5 to your oral presentation.
- Time yourself. Make sure your presentation does not exceed the time limit.
- Practice, practice, practice. With each round of practice, try to rely less and less on your notes.
- After you are satisfied with the organization, flow, and length, ask a friend to listen to your presentation. Ask him or her to evaluate your poise, posture, voice (clarity, volume, and rate), gestures and mannerisms, and interaction with the audience (eye contact, ability to recognize if the audience is following your talk).

It is natural to be nervous when you begin speaking to an audience, even an audience of your classmates. Adequate knowledge of the subject, good preparation, and sufficient rehearsal can all help to reduce your nervousness and enhance your self-assurance.

Delivery

The importance of the delivery cannot be overstated: listeners pay more attention to body language (50%) and voice (30%) than to the content (20%) (Fegert *et al.* 2002). Remember that you must establish a good rapport with the audience in order for your oral presentation to be successful. The following guidelines will help.

Presentation style

- Dress appropriately for the occasion.
- Use good posture. Good posture is equated with self-assurance, while slouching implies a lack of confidence.
- Be positive and enthusiastic about your subject.
- Try to maintain eye contact with some members of the audience. In the United States, eye contact is perceived as more personal, as though the speaker is having a conversation with individual listeners.
- Avoid distracting gestures and mannerisms such as pacing, fidgeting with the pointer, jingling the change in your pocket, and adjusting your hair and clothes.
- Speak clearly, at a rate that is neither too fast nor too slow, and make sure your voice carries to the back of the room. Avoid "um," "ah," "like," and other nervous sounds.
- Do not stand behind the computer monitor and read your presentation off the computer screen. Similarly, do not turn your back on the audience and read your presentation off the projection screen. Both practices distance you from your audience and make you seem unprepared.
- If you use notecards, number them so that if you drop them you can reassemble them quickly.
- Do not display a visual until you are ready to discuss it.
- When pointing to something on the projection screen, stand close enough to point with your outstretched arm or pointer (stick) or use a laser pointer.

- Tell the audience exactly what to look for. Interpret statistics and numbers so that they are meaningful to your listeners. If you are describing Figure 8.3, for example, "Beta-galactosidase production increased dramatically in only 10 minutes after adding the inducer" is likely to make a bigger impression on the audience than "From 0 to 10 minutes, the absorbance increased from 0.01 to 0.24."
- Point to the specific part of the visual that illustrates what you are describing. Avoid waving the pointer in the general direction of the visual.
- Do not block the projected image with your body.

Interacting with the audience

A unique benefit of communicating information orally is that you can adjust your presentation based on the feedback you receive from your listeners. Pay attention to the reaction of individuals in the audience. Does their posture or facial expression suggest interest, boredom, or confusion? If you perceive that the audience is not following you, ask a question or two, and adjust your delivery according to their response. Your willingness to customize your presentation to your audience will enhance your reputation and lead to greater listener satisfaction.

To keep listeners focused:

- Establish common ground immediately. Ask a question or describe a general phenomenon with which your audience is familiar.
- Incorporate humor if you can do so without being culturally insensitive or offensive.
- Make sure your introduction is well organized and proceeds logically from the general to the specific without any sidetracks.
- Alert your audience when you plan to change topics. Practice these transitions when you rehearse.
- Summarize the main points from time to time. Tell your listeners what you want them to remember.
- Use examples that are relevant to your audience. Keep in mind that older members of the audience may not be familiar with the latest viral videos on the Internet. Conversely, younger members of the audience may not know what TV shows were popular in the 1980s.
- End your presentation with a definite statement such as "In conclusion..." Don't fade out weakly with phrases such as "Well, that's about it."

Group presentations

If you are presenting with another person, your individual contributions should complement one another, not compete with each other. This requires good coordination and practice beforehand.

Fielding listener questions

Allow time for questions and discussion. Take questions as a compliment—your listeners were paying attention! Listen carefully to each question, repeat it so that everyone in the room can hear, and then give a brief, thoughtful response. If you don't understand the question, ask the listener to rephrase it. If you don't know the answer, say so. You might offer to find out the answer and follow up with the person later.

Signal the end of the question session with "We have time for one more question" or something similar.

Slide Decks as References

A **slide deck** is the group of slides that makes up a presentation. Besides being used as visual aids during the presentation, slide decks have the potential to be valuable references in the absence of the speaker. Your instructors probably post their lecture slide decks on a learning management system such as Blackboard or Moodle for you to review. Research collaborators share slide decks to avoid having to recreate images and reexplain basic concepts. In the corporate world, slide decks enable those who are unable to attend a meeting to stay informed, have access to supporting documentation, and edit the slides for other presentations.

When the slide decks are conceived primarily as references instead of visual aids for the audience, invariably too much text ends up on the slides. A solution to this problem is to use the notes pane for explanatory text, and design the slide so that the audience can focus on the take-home message. To open the notes pane in PowerPoint 2013, 2016, and PowerPoint for Mac 2016, click **Notes** at the bottom of the window. Type your talking points as bullets or running text; there is no limit to the length of the notes. Print out the notes pages or save them as a PDF:

- Windows: **File | Print | Print Layout | Notes Pages**
- Mac: **File | Print | Layout : Notes**

In this format, each page consists of a thumbnail of the slide along with the explanatory notes (Figure 8.5).

Notes can also be used as prompts for the speaker during the presentation. In **Presenter View**, the speaker can see the notes on the computer

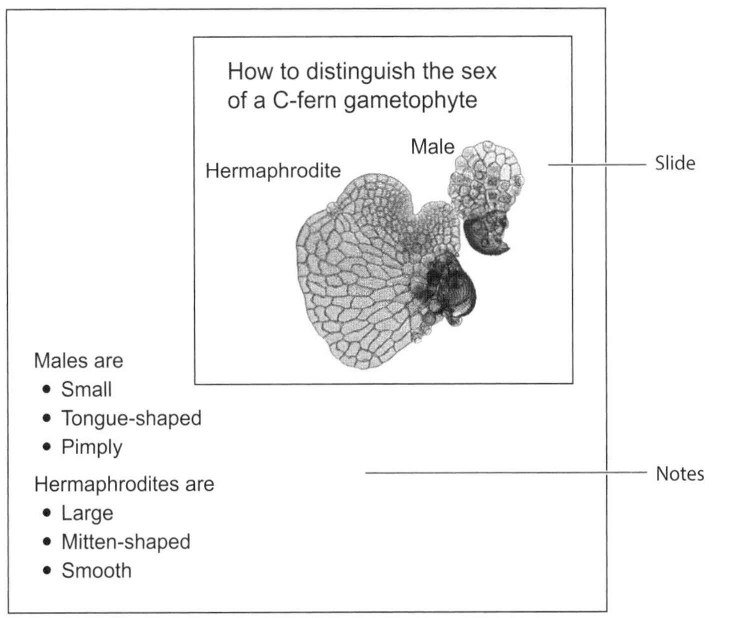

Figure 8.5 Slide decks saved as notes pages can be used as speaker notes during the presentation and as references before and after the presentation. Photo courtesy of Professor Mark Spiro, Bucknell University.

screen, but the audience sees only the current slide on the projection screen (Figure 8.6). **Presenter View** is available in PowerPoint 2013 and 2016 for Windows and PowerPoint for Mac 2016 (see pp. 267–269 in Appendix 3).

Slide decks can be shared in a variety of ways, as shown in Table 8.2.

Feedback

It takes not only practice, but also coaching, to become an effective speaker. When you make an oral presentation to your class, your instructor may ask your classmates to evaluate your delivery, organization, and visual aids, as well as the thoroughness of your research using a form such as the "Evaluation Form for Oral Presentations" available at <http://sites.sinauer.com/Knisely5E>.

When you are in the position of the listener, make the kinds of comments you would find helpful if you were the speaker. As you know, feedback is most likely to be appreciated when it is constructive, specific, and done in an atmosphere of mutual respect.

TABLE 8.2 Ways to share slide decks

Purpose	Format
To replay the presentation; to edit or reuse slides for other presentations	**File | Save as** PowerPoint Presentation (.ppt or .pptx)
To play the animations	**File | Save as** PowerPoint Presentation (.ppt or .pptx)
To provide **handouts for notetaking**. Print images of the slides (without notes) so that audience members can take notes during the talk.	Windows: **File | Print | Handouts | 3 Slides** Mac: **File | Print | Layout: Handouts (3 slides per page)** Select printer for paper copies or print to PDF if posting online
To provide **handouts with explanations**. Print a large thumbnail of the slide along with notes (one slide per page). This is a good option when there are a lot of notes.	Windows: **File | Print | Print Layout | Notes Pages** Mac: **File | Print | Layout: Notes (3 slides per page)** Select printer for paper copies or print to PDF if posting online
To provide **handouts with explanations economically**. Print up to 5 thumbnails of slides along with notes on one page. This is a good option when not all slides have notes or if notes are minimal.	Windows only: **File | Export | Create Handouts** Exports slides and notes to a Word document
To provide **handouts without explanations**. Print up to 6 thumbnails of slides without notes on one page.	Windows: **File | Print | Handouts |** and choose up to 6 slides per page Mac: **File | Print | Layout: Handouts (6 slides per page)**
To use as **speaker notes:** To view on the computer screen	Use **Presenter View**
To print out a paper copy: • For a lot of notes, see "Handouts with explanations."	Windows: **File | Print | Print Layout | Notes Pages** Mac: **File | Print | Layout: Notes**
• For minimal notes, see the economical handouts option.	Windows only: **File | Export | Create Handouts**

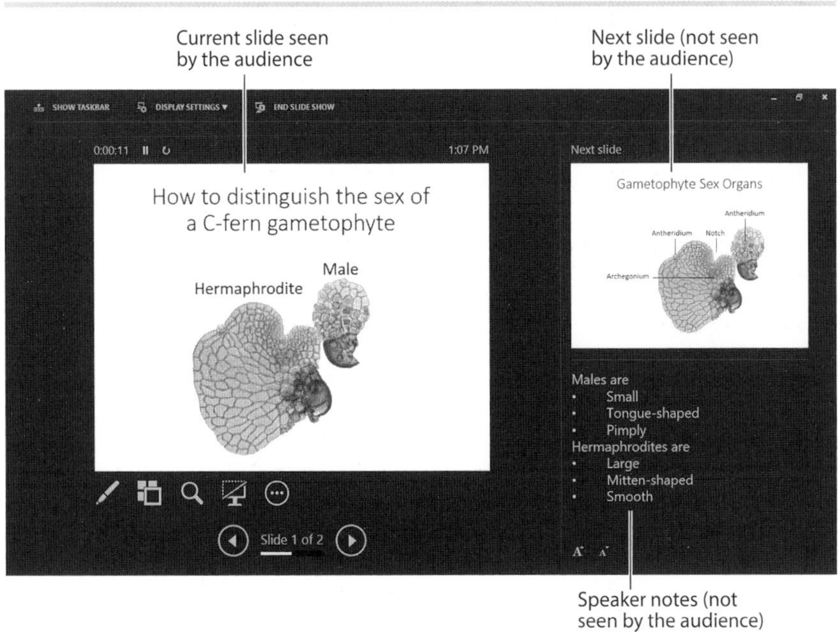

Figure 8.6 Presenter View in PowerPoint allows the speaker to view notes on the computer screen, eliminating the need to create text-heavy slides. Photo courtesy of Professor Mark Spiro, Bucknell University.

Go to the **COMPANION WEBSITE** • **sites.sinauer.com/knisely5e**
for samples, template files, and tutorial videos

Word Processing in Microsoft Word

Introduction

Microsoft Word 2013 and 2016 for Windows and Word for Mac 2016 have Ribbon interfaces. The **Ribbon** is a single strip that displays **commands** in task-oriented **groups** on a series of **tabs** (Figure A1.1). In Windows, additional commands in some of the groups can be accessed with the **dialog box launcher**, a diagonal arrow in the right corner of the group label. On Macs, additional commands are located on the **menu bar**. The **Quick Access Toolbar** comes with buttons for saving your file and undoing and redoing commands; you can also add buttons for tasks you perform frequently.

The Ribbon interface has characterized the past several versions of Word, so finding commands in the latest version should not be difficult. The main differences between Word 2013 and 2016 are cloud integration by default, additional tabs, Smart Lookup directly through the internet, the ability to convert PDFs to Word documents, and more flexibility for collaboration, similar to Google Docs.

This appendix will address commands frequently used when writing scientific papers. See also the video tutorials for Windows and Mac at <http://sites.sinauer.com/Knisely5E>. For all other word processing questions, please visit the Microsoft Office support website.

In this book, the nomenclature for the command sequence is as follows: **Ribbon tab | Group** (Windows only) | **Command button | Additional Commands** (if available). For example, in Windows, **Home | Font | Superscript** means "Select the **Home** tab on the Ribbon, and in the **Font** group, click the **Superscript** button." In the Mac OS, the corresponding command sequence is **Home | Superscript**.

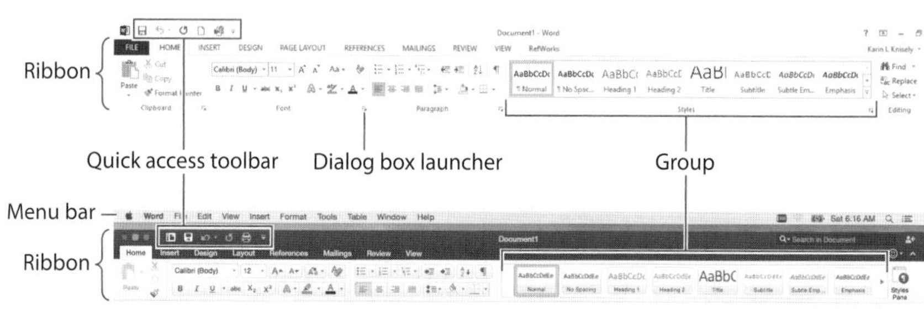

Figure A1.1 Screen display of command area in (A) Word 2013 and (B) Word for Mac 2016.

Good Housekeeping

Organizing your files in folders

You can expect to type at least one major paper and perhaps several minor writing assignments in every college course each semester. This amounts to a fair number of documents on your computer. One way to organize your files is to make individual course folders that contain subfolders for lecture notes, homework, and lab reports, for example. To create a new folder in Word when you are saving a file:

Mac

1. Click **File | Save As**. The **Save As** dialog box appears.
2. Click **Online Locations | +** to access **OneDrive** to save your file to Microsoft's cloud (you must have set up an account to use this option).
3. If you want to save the file to your computer or iCloud Drive, do *not* click **Online Locations**. Instead, click the down arrow next to the **Save As** box, and navigate to a preexisting folder or create a new folder.
4. To see the files you stored on iCloud Drive, click **Finder | iCloud Drive**.
5. To create a new folder, click the **New Folder** button. The **New Folder** dialog box appears.
6. Give the folder a short, descriptive name. Click **Create**.

To create a new folder in **Finder**:

1. Click **Finder**. Navigate to the desired location.
2. Click **File | New Folder** on the menu bar. Type a name for the new folder and press **Return**.

Windows

1. Click **File | Save As**. The **Save As** dialog box appears.
2. Click **OneDrive** to save to Microsoft's cloud (you must have set up an account to use this option). If you want to save the file to your computer, navigate to a preexisting folder or create a new folder.
3. To create a new folder, click the **New Folder** button. The **New Folder** dialog box appears.
4. Give the folder a short, descriptive name. Click **OK**.

To create a new folder in **Windows Explorer**:

1. Click **Start | File Explorer | Documents**. The Documents library dialog box appears.
2. Click the **New folder** button in the top left of the box. Type a name for the new folder and press **Enter**.

Accessing files and folders quickly

Mac Word makes recently used files readily accessible by displaying them when you click **File | Open Recent**. Frequently used files can also be pinned to the **Recent Documents** list:

1. Click **File | Open Recent | More**.
2. When you hold the cursor over a file, a thumbtack will appear on the right. Click the thumbtack to pin the file to the top of the list.
3. When you are finished working on the file, click **File | Open Recent | More**, and click the pinned file to remove it from the list.

Windows Word makes recently used files readily accessible by displaying them when you click **File | Open**. Frequently used files can also be pinned to the **Recent Documents** list:

1. Click **File | Open | Recent Documents**.
2. Right-click a file and select **Pin to list** from the menu.
3. The file will be pinned to the top of the list.

4. When you are finished working on the file, click **File | Open | Recent Documents**, right-click the pinned file, and select **Remove from list**.

To pin files or folders to **Quick access** in Windows Explorer:

1. Open a Windows Explorer dialog box by clicking **Start | File Explorer | Documents**.
2. Browse your documents library until you find the frequently used file (folder).
3. Hold down the left mouse button and drag the file (folder) to Quick access in the navigation pane on the left.

Naming your files

File names can include letters, numbers, underlines, and spaces as well as certain punctuation marks such as periods, commas, and hyphens. The following characters are not allowed: \, /, :, *, ?, ", <, >, and |.

In general, file names should be short and descriptive so that you can easily find the file on your computer. If you intend to share your file, however, consider the person you are sharing it with. When you send your lab report draft to your instructor for feedback, put your name and the topic in the file name (for example, Miller_lacoperon). Although "biolab1" may seem unambiguous to you, it is not exactly informative for your biology professor!

If you have forgotten the exact file name and location, you can search for it. On a Mac, use **Finder**. In Windows, click **Start | File Explorer | Computer | Local Disk (C:)** (or **Documents** if you are able to narrow down the location).

Saving your documents

When you write the first draft of a paper by hand, you have tangible evidence that you have done the work. When you type something on the computer, however, your work is unprotected until you save it. That means that if the power goes off or the computer crashes before you save the file, you have to start again from scratch. Hopefully it won't take the loss of a night's work to convince you to save your work early and often. Don't wait until you've typed a whole page; save your file after the first sentence. Then continue to save it often, especially when the content and format are complicated. Think about how long you would need to retype the text if it were lost and if you can afford to spend that much time redoing it.

Word automatically saves your file at certain intervals. You can adjust the settings in Windows by clicking **File | Options | Save | Save Auto-**

Recover information every __ minutes. In the Mac OS, click **Word | Preferences** and then **Output and Sharing | Save**. Enter a time for **Save every __ minutes**. It's also a good idea to save the file manually from time to time by clicking 🖫 on the **Quick Access Toolbar**.

Backing up your files

It should be obvious that your backup files should not be saved on your computer's hard drive. Table A1.1 lists some offline and online backup options along with their advantages and disadvantages. In terms of offline options, USB flash drives are probably your best bet for compactness and convenience, but external hard drives come with software that lets you schedule automatic backups. With an account and an internet connection, you can take advantage of any number of online options. These include cloud services that store your files virtually, saving files to your organization's server, and even emailing files to yourself.

The most important thing about backing up your files is to have a plan. The odds are, unfortunately, that you will have a computer malfunction at least once every 3 years. For that reason alone:

- Schedule regular, automatic backups for your hard drive (that way you won't forget).
- Have a second backup option for important files that you are currently working on (especially projects with tight deadlines).
- Make sure your backup options don't run out of storage space. As a rough guide, a 10-page Word document takes up about 0.2 MB. In comparison, a picture taken with your smartphone is typically around 2-5 MB, and each of your music files can be as large as 8-10 MB.

The bottom line is: Protect your valuable files with a reliable backup system.

AutoCorrect

AutoCorrect is useful for more than just correcting spelling mistakes. Increase your efficiency by programming AutoCorrect with a simple keystroke combination to replace an expression that takes a long time to type. You must choose the simple keystroke combination judiciously, however, because every time you type these keystrokes, Word will replace them with what you programmed in AutoCorrect. One "trick" is to precede the keystroke combination with either a comma or semicolon (, or ;). Since punctuation marks are typically followed by a space and not a letter, adding one before your "code letter(s)" creates a unique combination.

TABLE A1.1 Possible options for backing up your electronic files

Backup Method	Capacity	Benefits	Drawbacks
USB flash drive	Up to 1 TB	Portable Can be encrypted No internet connection needed	Expensive per GB compared to hard drives Data lost if device is lost
SD/microSD card	Up to 512 GB	Portable Can be read from mobile devices No internet connection needed	Expensive per GB compared to hard drives Data lost if device is lost
External hard drive	Up to 8 TB	Portable Inexpensive Can be encrypted No internet connection needed Can perform automatic backups	Larger physical size compared to flash drives or SD cards Data lost if device is lost
Internal server with hard disk drives	Depends on organization	No subscription needed Files can be shared or kept private Can perform automatic backups	In-house networking required Internet required to access files remotely

Long words

Let's say "beta-galactosidase" is a word you have to type frequently. Choose ";bg" to designate "beta-galactosidase." To program AutoCorrect for this entry:

Mac

1. Type "beta-galactosidase" and select it.
2. On the menu bar, click **Tools | AutoCorrect** or **Word | Preferences | Authoring and Proofing Tools | AutoCorrect.**
3. Type ";bg" in the **Replace** text box. You'll see that "beta-galactosidase" is automatically entered in the **With** text box.
4. Click the **Add** button.

TABLE A1.1 *(continued)*

Backup Method	Capacity	Benefits	Drawbacks
Internal server with solid state drives	Depends on organization	Fastest bulk data storage No subscription needed Files can be shared or kept private Can perform automatic backups	Expensive In-house networking required Internet required to access files remotely
Cloud storage Google Drive Microsoft OneDrive Apple iCloud Box Dropbox Mozy SugarSync Mega	Infinite with subscription	2–50 GB of free storage Can perform automatic backups Can be accessed from any computer or mobile device with internet Files can be shared with collaborators	Subscription needed for larger storage tiers Prices increase with storage needs Internet connection required
Send emails to yourself	Typically 15 GB, potentially unlimited with organization email	Free Can be accessed from any computer with internet	Poor file organization Internet required 10-25 MB file size limit

Source: B. Knisely, personal communication, 18 December 2016.

5. Click the **red X** button to exit.
6. Next time you have to write "beta-galactosidase," simply type ";bg."

Windows

1. Type "beta-galactosidase" and select it.
2. Select **File | Options | Proofing | AutoCorrect Options**.
3. Type ";bg" in the **Replace** text box. You'll see that "beta-galactosidase" is automatically entered in the **With** text box.
4. Click the **Add** button.
5. Click **OK** to exit the **AutoCorrect** dialog box.
6. Click **OK** to exit the **Word Options** box.
7. Next time you have to write "beta-galactosidase," simply type ";bg."

Expressions with sub- or superscripts

Mac

1. First type the expression without sub- or superscripts (e.g., Vmax).
2. Select the characters to be sub- or superscripted, and then format them by clicking the appropriate button on the **Home** tab of the Ribbon (the expression then becomes V_{max}).
3. Select the entire expression and click **Tools | AutoCorrect**.
4. On the **AutoCorrect** tab, you will notice "Vmax" already entered in the **With** text box. Click the **Formatted text** option button to add the subscripting. The expression then becomes V_{max}.
5. Type "Vmax" in the **Replace** text box.
6. Click the **Add** button.
7. Click the **red X** button to exit.

Windows

1. First type the expression without sub- or superscripts (e.g., Vmax).
2. Select the characters to be sub- or superscripted, and then format them by clicking the appropriate button on **Home | Font** (the expression then becomes V_{max}).
3. Select the entire expression and click **File | Options | Proofing | AutoCorrect Options**.
4. On the **AutoCorrect** tab, you will notice "Vmax" already entered in the **With** text box. Click the **Formatted text** option button to add the subscripting. The expression then becomes V_{max}.
5. Type "Vmax" in the **Replace** text box.
6. Click the **Add** button.
7. Click **OK** to exit the **AutoCorrect** dialog box.
8. Click **OK** to exit the **Word Options** box.

Italicize scientific names of organisms automatically

By convention, scientific names of organisms are always italicized. When you type the name of the organism, capitalize genus but write species in lower case (e.g., *Homo sapiens*). To automate italicization, follow these steps:

Mac

1. First type the scientific name of the organism in your text (e.g., Aphanizomenon flos-aquae).
2. Italicize the name by selecting it and then clicking the **Italic** button (***I***) on the **Home** tab. The text then becomes: *Aphanizomenon flos-aquae.*
3. With the name still selected, click **Tools | AutoCorrect**.
4. On the **AutoCorrect** tab, you will notice Aphanizomenon flos-aquae (without italics) already entered in the **With** text box. Click the **Formatted text** option button to italicize it.
5. Enter a shorter version of the name in the **Replace** text box, for example "aphan."
6. Click the **Add** button.
7. Click the **red X** button to exit.

Windows

1. First type the scientific name of the organism in your text (e.g., Aphanizomenon flos-aquae).
2. Italicize the name by selecting it and then clicking the **Italic** button (***I***) on **Home | Font**. The text then becomes: *Aphanizomenon flos-aquae.*
3. With the name still selected, click **File | Options | Proofing | AutoCorrect Options**.
4. On the **AutoCorrect** tab, you will notice Aphanizomenon flos-aquae (without italics) already entered in the **With** text box. Click the **Formatted text** option button to italicize it.
5. Enter a shorter version of the name in the **Replace** text box, for example "aphan."
6. Click the **Add** button.
7. Click **OK** to exit the **AutoCorrect** dialog box.
8. Click **OK** to exit the **Word Options** box.

Endnotes in Citation-Sequence System

The most useful feature on the References tab, at least with regard to scientific papers, is the **Insert Endnote** (not footnote!) command for in-text references in the Citation-Sequence (C-S) system. To insert superscripted endnotes, type the sentence containing information that requires a reference. End the sentence with a period.

Mac

1. Then click **References | Insert Endnote**.
2. Type the full reference in proper format (see "The Citation-Sequence System" in Chapter 4).
3. Click **References | Show Notes** to exit the endnote and return to the text.
4. To change the endnote number style from Roman to Arabic, click **Insert | Footnotes** on the menu bar and select Arabic number format from the drop-down menu.

Windows

1. Then click **References | Footnotes | Insert Endnote**.
2. Type the full reference in proper format (see "The Citation-Sequence System" in Chapter 4).
3. Click **References | Footnotes | Show Notes** to exit the endnote and return to the text.
4. To change the endnote number style from Roman to Arabic, click the Footnotes dialog box launcher and select Arabic number format from the drop-down menu.

Word updates endnotes automatically so that they appear sequentially. To see an endnote, point the mouse at the superscripted number. To edit an endnote, double-click the number. To delete an endnote, highlight the superscripted number and press **Delete**. The other endnote numbers will be renumbered automatically.

Equations

Mac

To insert a common mathematical equation or make your own, click **Insert | Equation** to display the **Equation** tab. Scan the built-in equations in the drop-down menu for the **Equation** button on the left side of the tab. If this menu does not contain the equation you'd like to write, choose a form to modify from the **Structures** group. Two commonly used structures are **Fraction** (numerator over denominator terms) and **Script** (subscripts and superscripts). Fill in numbers or select Greek letters, arrows, and mathematical operators from the **Math Symbols** group.

Windows

To insert a common mathematical equation or make your own, click **Insert | Equation** to display the **Equation Tools Design** tab. Scan the built-in equations in the drop-down menu for the **Equation** button on the left side of the tab. If this menu does not contain the equation you'd like to write, choose a form to modify from the **Structures** group. Two commonly used structures are **Fraction** (numerator over denominator terms) and **Script** (subscripts and superscripts). Fill in numbers or select Greek letters, arrows, and mathematical operators from the 8 symbol sets that appear after clicking the **More** arrow.

Feedback Using Comments and Track Changes

It may not always be possible for you and your collaborator (or reviewer) to find a common time to meet to go over your paper. Email makes the review process more convenient. You can send your paper to a reviewer in electronic format as an attached file, and your reviewer can send it back to you after making comments or revisions directly in the document.

Comments are frequently used to exchange ideas or ask questions when collaborating on a paper. You are most likely to add a comment, read someone else's comment and respond with a comment of your own, or delete one or more comments.

Some reviewers will make edits directly in the text using Track Changes. When this command is turned on, any changes made in the document will be colored and underlined. Any text that is deleted will be shown in a balloon to the right. Each reviewer gets his or her own color so you can identify the changes made by multiple reviewers. In addition, a vertical line in the left margin alerts you to changes in your document. Clicking this line toggles between **All Markup** and **Simple Markup** modes.

Mac

Add a comment

1. Select the text you want to comment on.
2. Click **Review | New Comment**.
3. Type your comment in the box. Click outside of the comment area to return to the document.

Review and then delete comments When there are no comments in a document, the **Delete**, **Previous**, and **Next** buttons are grayed out. When there are comments, these commands become available. After you have read and, if appropriate, taken action on all of the comments, delete them. To delete comments one at a time, click the comment balloon and then click **Delete**. Use the **Next** button in the Comment group to go to the next comment. To delete all of the comments at once, click the down arrow next to **Delete** and then **Delete All Comments in Document**.

Track changes

1. To turn on tracking, click **Review | Track Changes**. The button is green when tracking is turned on. To turn off tracking, click the button to display **Off**.
2. To review the edits made by others, click the **Next** button in the Reviewing group. Use the buttons for **Accept** or **Reject** or simply click **Next** to read the suggestion without taking any action.
3. Before you turn in your lab report, make sure all edits have been addressed (by accepting or rejecting them) and tracking is turned off. *Please note that accepting/rejecting changes does not remove the comments.* Comments must be deleted separately, as described in the previous section.

Windows

Add a comment

1. Select the text you want to comment on.
2. Click **Review | Comments | New Comment**.
3. Type your comment in the box. Click outside of the comment area to return to the document.

Review and then delete comments When there are no comments in a document, the **Delete**, **Previous**, and **Next** buttons in the Comments group are grayed out. When there are comments, these commands become available. After you have read and, if appropriate, taken action on all of the comments, delete them with **Review | Comments | Delete | Delete All Comments in Document**. To delete comments one at a time, click the balloon and select **Delete Comment**.

Track changes

1. To turn on tracking, click **Review | Tracking | Track Changes**. The button is blue when tracking is turned on. To turn off tracking, click the button so that it turns white.
2. To review the edits made by others, click the **Next** button in the Changes group. Use the buttons for **Accept** or **Reject** or simply click **Next** to read the suggestion without taking any action.

3. *Please note that accepting/rejecting changes does not remove the comments.* Comments must be deleted separately, as described in the previous section.

Document inspector Before you turn in your lab report, check that you've removed all comments and tracked changes. Click **File | Info | Check for Issues | Inspect Document**. Rather than selecting **Remove All**, click **Close**. Find the stray revision marks by clicking **Review | Changes | Next** and take action, if necessary, and delete or reject them.

Format Painter

This paintbrush is great for copying and pasting format, whether it is the typeface or font size of a character or word, the indentation of a paragraph, or the position of tab stops. It's also handy for resuming numbering in a numbered list. It works with character, word, and paragraph formatting, but not page formatting.

To copy the format of a word, select the word, and click **Home |** (Mac) or **Home | Clipboard | Format Painter** (Windows). Drag the cursor (which has changed to a paintbrush next to an I-beam) over the text you want to format. To copy paragraph formatting (including tab stops), click inside the paragraph, but don't select anything. Then click the paintbrush and drag the mouse over the paragraph to which you want to apply the format.

To apply the format to multiple blocks of text, click the formatted text and then double-click the paintbrush (**Format Painter**). Apply the formatting to the selected blocks of text. Click the paintbrush again to turn off this function.

Formatting Documents

An electronic file called "Biology Lab Report Template," available at <http://sites.sinauer.com/Knisely5E> is formatted according to these specifications.

Margins

Set 1.25″ left and right and 1″ top and bottom margins to give your instructor room to make comments.

Mac Click **Layout | Margins**.

Windows Click **Page Layout | Page Setup | Margins**.

Paragraphs

Mac The **Paragraph** dialog box is opened by clicking **Format | Paragraph** on the menu bar.

First line indent is used to denote the beginning of a new paragraph. Under **Indentation | Special**, select **First line** from the drop-down menu and then specify **By: 0.5"** (or your preference).

Hanging indent is used for listing full references. The first line of each reference begins on the left margin and the subsequent lines are indented. First type each reference so that it is aligned on the left margin and ends with a hard return ¶. When you are finished, select all of the references. Under **Indentation | Special**, select **Hanging** from the drop-down menu and then specify **By: 0.25"** (or your preference).

Line spacing for all lines of text is adjusted with the **Line spacing** button. The default spacing is **Single**; your instructor may ask you to change the spacing to **Double** in order to have a little extra space to write comments.

Spacing between paragraphs. As an alternative to indenting the first line, paragraphs can be separated with a blank line. To automatically add extra space after each paragraph, thereby saving you the trouble of pressing the **Return** key twice, put the insertion pointer anywhere in the paragraph. In the **Paragraph** dialog box, enter **12 pt** in the **Spacing | After** box. This spacing corresponds to one blank line when using a 12 pt font size for your paper.

Line and page breaks. This feature is handy for preventing section headings from being separated from the body due to a natural page break. Select the section heading (for example, "Materials and Methods"). Then open the **Paragraph** dialog box and click the second tab called **Line and Page Breaks**. Under **Pagination**, check the box next to **Keep with next** to keep the heading together with the body. To keep all the lines of a paragraph together on the same page, check **Keep lines together**.

Windows The **Paragraph** dialog box can be accessed from both the **Home** and **Page Layout** tabs. Clicking the diagonal arrow on the **Paragraph** group label launches a dialog box where you can format the following paragraph attributes.

First line indent is used to denote the beginning of a new paragraph. Next to **Indentation | Special**, select **First line** from the drop-down menu and then specify **By: 0.5"** (or your preference). If you use this method, change the default paragraph spacing to 0. See "Paragraph spacing" below.

Hanging indent is used for listing full references. The first line of each reference begins on the left margin and the subsequent lines are indented. First type each reference so that it is aligned on the left margin and ends with a hard return ¶. When you are finished, select all of the references and click **Home | Paragraph** dialog box launcher. Next to **Indentation | Special**, select **Hanging** from the drop-down menu and then specify **By: 0.25"** (or your preference).

Line spacing for all lines of text. The default spacing in Word 2013 is **Multiple at 1.08 lines**; your instructor may ask you to change the spacing to **Double** in order to have a little extra space to write comments. There is also a **Line spacing** command button under **Home | Paragraph** where you can make this change.

Paragraph spacing. As an alternative to indenting the first line, paragraphs can be separated with a blank line. The default after-paragraph spacing in Word 2013 is 8 pt. To increase this space to the height of a blank line, enter **12 pt** in the **Spacing | After** box. If you like the space between paragraphs, then it is not necessary to use the first line indent method.

Line and page breaks. This feature is handy for preventing section headings from being separated from the body due to a natural page break. Select the section heading (for example, "Materials and Methods"). Then open the **Paragraph** dialog box and click the second tab called **Line and Page Breaks**. Under **Pagination**, check the box next to **Keep with next** to keep the heading together with the body. To keep all the lines of a paragraph together on the same page, check **Keep lines together**.

Page numbers

Page numbers make it easier for you to assemble the pages of your document in the correct order.

Mac

1. To insert a page number, click **Insert | Page Number** from either the menu bar or the Ribbon, and then select a position for the numbers.
2. If you do not want the number to appear on the first page, uncheck **Show number on first page**.
3. To start numbering with a custom number (for example, if the documents are chapters in a book), click **Insert | Page Number**, and enter a number under **Page numbering | Start at: __**.

Windows

1. To insert a page number, click **Insert | Header & Footer | Page Number**, and then select a position for the numbers.
2. If you do not want the number to appear on the first page, click **Different first page**.

3. To start numbering with a custom number (for example, if the documents are chapters in a book), click **Insert | Header & Footer | Page Number | Format Page Numbers**, and enter a number under **Page numbering | Start at: __**.

Proofreading Your Documents

Before you waste reams of paper printing out drafts of your document, have Word check spelling and grammar. Use your eyes to look over format on-screen before proofreading the printed document.

Spelling and grammar

Word gives you visual indicators to alert you to possible spelling and grammar mistakes. Words that are possibly misspelled are underlined with a wavy red line. Phrases that may contain a grammatical error or an extra space are underlined with a wavy green line in Mac, a wavy blue line in Windows. Do not ignore these visual cues!

Mac To deal with a word underlined in red, **control-click** it. A pop-up menu appears with commands and suggestions for replacements. **Ignore All** applies only to the current document. **Add to Dictionary** applies to all future documents. It makes sense to add scientific terminology to Word's dictionary after you consult your textbook or laboratory manual to confirm the correct spelling. After making a selection on the pop-up menu, the wavy red underline is deleted and the word is ignored in the manual, systematic spelling and grammar check. Similarly, to deal with a possible grammatical error underlined in green (including extra spaces between words), **control-click** it to accept or ignore Word's suggestions.

Windows To deal with expressions underlined in red or blue, right-click them. A pop-up menu appears with commands and suggestions for replacements. The options are the same as for Mac above.

Prevent section headings from separating from their body

Research articles are divided into sections: Abstract, Introduction, Materials and Methods, Results, Discussion, References, and Acknowledgments. Each section begins with a heading, on a separate line, followed by the body. When you check the format of your paper, make sure the heading is not cut off from the body of the section.

To prevent heading-body separation problems, use one of these options:

- See "Line and page breaks" in the Paragraphs section.

- Insert a hard page break to the left of the heading to force the heading onto the next page with the body. On a Mac, press **⌘ (command)+Enter**. In Windows, press **Ctrl+Enter**.

Prevent figures and tables from separating from their caption

See "Prevent section headings from separating from their body."

Errant blank pages

To delete blank pages in the middle of a document, go to the blank page, click it, and display the hidden symbols. For Mac, click **Home | ¶ (Show all nonprinting characters)**. For Windows, click **Home | Paragraph | Show/Hide ¶**. Delete spaces and paragraph symbols until the page is gone.

Similarly, if the blank page is at the end, navigate to the end of the document and remove the hidden symbols to delete the extra page. To navigate to the end of the document on a MacBook keyboard, press **⌘+fn+Right arrow**. On a laptop PC, press **Ctrl+Fn+End**.

Finally, print a hard copy

When you are confident that you've found and corrected all mistakes on-screen, print a hardcopy and proofread your paper again. Some mistakes are more easily identified on paper, and it is always better for you, rather than your instructor, to find them.

Sub- and superscripts

Expressions with sub- or superscripting are common in the natural sciences and their formatting makes them readily identifiable to members of the scientific community. Therefore, it would be incorrect to write sub- or superscripted characters on the same line as the rest of the text. Similarly, when using scientific notation, exponents are always superscripted. It is not acceptable to use a caret (^) to designate superscript or an uppercase E to represent an exponent.

RIGHT: 2×10^{-3} (exponent is superscripted)

WRONG: 2 × 10–3 or 2 × 10^–3 or 2 × 10E–3

If you frequently have to type expressions with sub- and superscripts, save time by programming them in AutoCorrect (see p. 196 in this appendix).

Mac

To superscript or subscript text, first type the expression without formatting. Then highlight the character(s) to be sub- or superscripted and click the appropriate button on the **Home** tab of the Ribbon. Alternatively, you can apply and remove sub- and superscripting using keyboard commands ("hot keys"). Select the character(s), and then hold down **⌘ (command)** while pressing = for subscript or hold down both **⌘** and **Shift** while pressing = for superscript.

Windows

To superscript or subscript text, first type the expression without formatting. Then highlight the character(s) to be sub- or superscripted and click the appropriate button under **Home | Font**. Alternatively, you can apply and remove sub- and superscripting using keyboard commands ("hot keys"). Select the character(s), and then hold down **Ctrl** while pressing = for subscript or hold down both **Ctrl** and **Shift** while pressing = for superscript.

Symbols

Mac

To insert Greek letters and mathematical symbols in running text, click **Insert | Advanced Symbol** and select the desired symbol.

Shortcut keys for symbols If you use a particular Greek letter or mathematical symbol frequently, you can make a keyboard shortcut to save time. Let's use the degree sign as an example.

1. After clicking this symbol in the **Symbol** dialog box (Unicode character F0B0), click the **Keyboard Shortcut** button at the bottom of the box.
2. In the **Customize Keyboard** dialog box, define a combination of keystrokes using **Control**, **⌘**, **Control+Shift**, or **Control+⌘** plus some other character. Because the degree sign looks like a lower case *o*, let's use **Control+o**. To define **Control+o** as the keyboard shortcut for the degree sign, hold down **Control** while pressing **o** in the **Press new shortcut key** box.
3. Word notifies you that this combination is unassigned or that it has already been assigned to another command. (Note: Even though you typed **Control** and lowercase **o**, the shortcut appears as Control and uppercase O in the box. If you had typed **Control** and uppercase **O**, this would have appeared as Control+Shift+O.)

4. Next time you have to write the symbols for "degrees Celsius," simply type **Control+o** followed by uppercase **C** to get: °C.

If the shortcut has already been assigned to a different command, assigning a symbol will automatically overwrite the original command. In other words, the same keystroke combination will insert the symbol instead of carrying out the original command. Do not overwrite the original command if this is one you use often!

AutoCorrect for symbols Another way to simplify inserting symbols is to program AutoCorrect. Once again, let's use the degree sign as an example.

1. After inserting this symbol in your document, select it and click **Tools | AutoCorrect** on the menu bar.
2. The ° symbol will already be entered in the **With** text box and the **Formatted text** option box will be selected. Type a unique combination of characters in the **Replace** text box. Because you may not want to replace the word "degrees" with the degree symbol (°) every time, use the trick of preceding the word with either a comma or semicolon (, or ;). Type ";degrees" in the **Replace** box. Click the **Add** button, and click the **red X** button to exit.
3. Next time you want to insert a degree sign, simply type ";degrees".

Windows

To insert Greek letters and mathematical symbols in running text, click **Insert | Symbols | Symbol | More Symbols**.

Shortcut keys for symbols If you use a particular Greek letter or mathematical symbol frequently, you can make a keyboard shortcut to save time. Let's use the degree sign as an example.

1. After clicking this symbol in the **Symbol** dialog box (Unicode character F0B0), click the **Keyboard Shortcut** button at the bottom of the box.
2. In the **Customize Keyboard** dialog box, define a combination of keystrokes using **Ctrl**, **Alt**, **Ctrl+Shift**, or **Ctrl+Alt** plus some other character. Because the degree sign looks like a lower case *o*, let's use **Control+o**. To define **Control+o** as the keyboard shortcut for the degree sign, hold down **Control** while pressing **o** in the **Press new shortcut key** box.
3. Word notifies you that this combination is unassigned or that it has already been assigned to another command. (Note: Even though you typed **Control** and lowercase **o**, the shortcut appears

as Control and uppercase O in the box. If you had typed **Control** and uppercase **O**, this would have appeared as Control+Shift+O.)

4. Next time you have to write the symbols for "degrees Celsius," simply type **Control+o** followed by uppercase **C** to get: °C.

If the shortcut has already been assigned to a different command, assigning a symbol will automatically overwrite the original command. In other words, the same keystroke combination will insert the symbol instead of carrying out the original command. Do not overwrite the original command if this is one you use often!

AutoCorrect for symbols Another way to simplify inserting symbols is to program AutoCorrect. Once again, let's use the degree sign as an example.

1. Open the **Symbol** dialog box, click the degree sign, and then click the **AutoCorrect...** button at the bottom of the box.
2. The ° symbol will already be entered in the **With** text box and the **Formatted text** option box will be selected. Type a unique combination of characters in the **Replace** text box. Because you may not want to replace the word "degrees" with the degree symbol (°) every time, use the trick of preceding the word with either a comma or semicolon (, or ;). Type ";degrees" in the **Replace** box. Click the **Add** button, click **OK** to exit the **AutoCorrect** dialog box, and finally click **Close** to exit the **Symbol** box.
3. Next time you want to insert a degree sign, simply type ";degrees".

Tables

Mac

Creating a table

1. Position the insertion pointer where you want to insert the table.
2. Click **Insert | Table**. Highlight the desired number of columns and rows. A blank table appears with the cursor in the first cell. To change the table format, use the **Table Design** and **Table Layout** tabs on the Ribbon. These tabs only appear when the insertion pointer is inside the table.
3. To insert or delete columns or rows or to merge or split cells, use the command buttons on the **Table Layout** tab.

Formatting text within tables

1. To apply formatting to adjacent cells, first select the cells by clicking the first cell in the range, holding down the **Shift** key, and then clicking the last cell.
2. With the cells still selected, apply, for example, boldface to the text by clicking **Home | B (Bold)**.

Viewing gridlines It is easier to enter data in a table when the gridlines are displayed. Position the insertion pointer inside the table and click **Table Layout | View Gridlines**. Gridlines are not printed.

Lines in a table By convention, tables in scientific papers do not have vertical lines to separate the columns, and horizontal lines are used only to separate the table caption from the column headings, the headings from the data, and the data from any footnotes (notice the format of the tables in this book).

1. Select all of the cells. Click **Table Design | Borders** and select **No Border**. You will still see the gridlines if they are selected (see "Viewing gridlines"), but gridlines are not printed.
2. Select the cells in the first row. Click **Table Design | Borders** and check **Bottom Border** and **Top Border**.
3. Select the cells in the last row. Click **Table Design | Borders** and check **Bottom Border**.

Navigating in tables

1. To jump from one cell to an adjacent one, use the arrow keys.
2. To move forward across the row, use the **Tab** key. Note: If you press **Tab** when you are in the last cell of the table, Word adds another row to the table.
3. To align text on a tab stop in a table cell, press **Ctrl+Tab**.

Anchoring images within tables Getting images to line up horizontally and vertically on a page can be tricky. It's possible to make multiple columns using **Table Layout | Columns** or to adjust the position of each picture on the **Picture Format** tab, but quite often the images become misaligned when the document is edited. To make images stay put, create a table. Select the desired layout (for example, 3 columns × 1 row), remove all borders, size the images so they fit into the cells, and then insert the images into the table.

Repeating header rows When a table extends over multiple pages, it is convenient to have Word automatically repeat the header row at the top of each page. To activate this command, click anywhere in the header row and then select **Table Layout | Repeat Header Rows**.

Windows

Creating a table

1. Position the insertion pointer where you want to insert the table.
2. Click **Insert | Tables | Table**. Highlight the desired number of columns and rows.

A blank table appears with the cursor in the first cell. To change the table format, use the **Table Tools Design** and **Table Tools Layout** tabs. These tabs only appear when the insertion pointer is inside the table.

1. To insert or delete columns or rows or to merge or split cells, use the command buttons on the **Table Tools Layout** tab.

Formatting text within tables

1. To apply formatting to adjacent cells, first select the cells by clicking the first cell in the range, holding down the **Shift** key, and then clicking the last cell.
2. With the cells still selected, apply, for example, boldface to the text by clicking **Home | Font | B (Bold)**.

Viewing gridlines It is easier to enter data in a table when the gridlines are displayed. Position the insertion pointer inside the table and click **Table Tools Layout | Table | View Gridlines**. Gridlines are not printed.

Lines in a table By convention, tables in scientific papers do not have vertical lines to separate the columns, and horizontal lines are used only to separate the table caption from the column headings, the headings from the data, and the data from any footnotes (notice the format of the tables in this book).

1. Select all of the cells. Click **Table Tools Design | Borders | Borders** and select **No Border**. You will still see the gridlines if they are selected (see "Viewing gridlines"), but gridlines are not printed.
2. Select the cells in the first row. Click **Table Tools Design | Borders | Borders** and check **Bottom Border** and **Top Border**.

3. Select the cells in the last row. Click **Table Tools Design | Borders | Borders** and check **Bottom Border**.

Navigating in tables

1. To jump from one cell to an adjacent one, use the arrow keys.
2. To move forward across the row, use the **Tab** key. Note: If you press **Tab** when you are in the last cell of the table, Word adds another row to the table.
3. To align text on a tab stop in a table cell, press **Ctrl+Tab**.

Anchoring images within tables Getting images to line up horizontally and vertically on a page can be tricky. It's possible to make multiple columns using **Page Layout | Page Setup | Columns** or to adjust the position of each picture on the **Picture Tools Format** tab, but quite often the images become misaligned when the document is edited. To make images stay put, create a table. Select the desired layout (for example, 3 columns × 1 row), remove all borders, size the images so they fit into the cells, and then paste the images into the table.

Repeating header rows When a table extends over multiple pages, it is convenient to have Word automatically repeat the header row at the top of each page. To activate this command, click anywhere in the header row and then select **Table Tools Layout | Data | Repeat Header Rows**.

Go to the **COMPANION WEBSITE** • **sites.sinauer.com/knisely5e**
for samples, template files, and tutorial videos

Making Graphs in Microsoft Excel

Introduction

Microsoft Excel 2013 and 2016 for Windows and Excel for Mac 2016 have Ribbon interfaces. The **Ribbon** is a single strip that displays **commands** in task-oriented **groups** on a series of **tabs** (Figure A2.1). In Windows, additional commands in some of the groups can be accessed with the **dialog box launcher**, a diagonal arrow in the right corner of the group label. In the Mac OS, additional commands are located on the **menu bar**. The **Quick Access Toolbar** comes with buttons for saving your file and undoing and redoing commands; you can also add buttons for tasks you perform frequently. The Ribbon interface has characterized the past several versions of Word, so finding commands in the latest version should not be difficult.

This appendix has three parts. The "Formulas" section introduces you to writing and applying formulas to save you time on data analysis. "Formatting the Spreadsheet" explains how to wrap text, change the number of displayed decimal places, change cell attributes, sort data, format and print the spreadsheet itself, change the view, and add and rename worksheets. The bulk of this appendix gives detailed instructions for Windows and Mac on how to make different types of graphs and format them according to Council of Science Editors (CSE) standards. See also the video tutorials for Windows and Mac at <http://sites.sinauer.com/Knisely5E>. For all other Excel questions, please visit the Microsoft Office support website.

In this book, the nomenclature for the command sequence is as follows:

> **Ribbon tab | Group** (Windows only) **| Command button | Additional Commands (if available).**

For example, to make an XY graph in Windows, **Insert | Charts | Insert Scatter (X, Y) | Scatter with Straight Lines and Markers** means "Select the **Insert** tab on the Ribbon, and in the **Charts** group, click the down arrow on the **Insert Scatter (X, Y)** button, and select the **Scatter with Straight Lines and Markers** option." In the Mac OS, the corre-

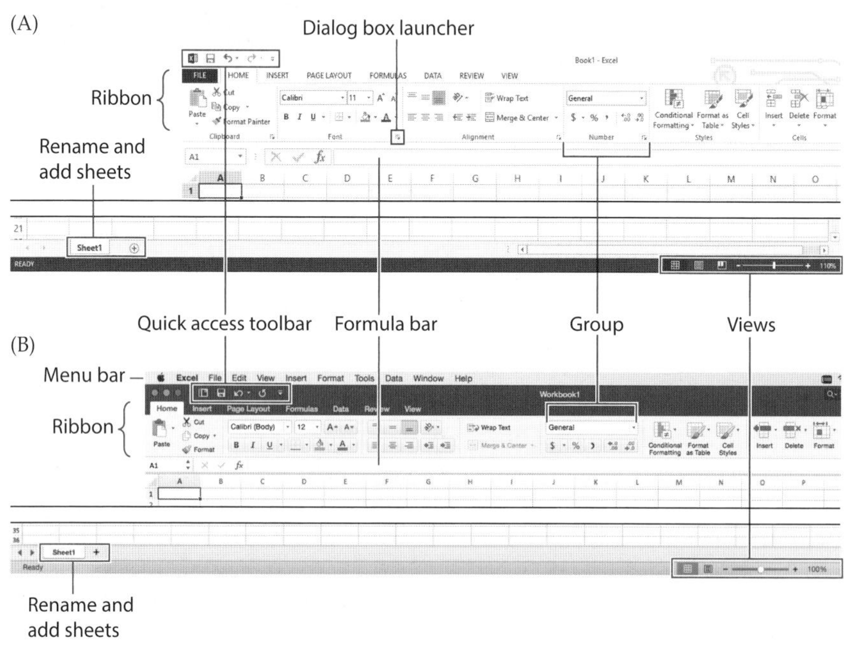

Figure A2.1 Screen display of part of the command area in (A) an Excel 2013 worksheet, and (B) an Excel for Mac 2016 worksheet. The status bar at the bottom of the screen has commands for changing views and adding and renaming worksheets.

sponding command sequence is **Insert | X Y (Scatter) | Scatter with Straight Lines and Markers**.

Handling computer files

The section "Good Housekeeping" in Appendix 1 applies equally to Word documents and Excel workbooks. Read over this section to develop good habits for naming, organizing, and backing up computer files.

Formulas in Excel

Excel is a popular spreadsheet program in the business world, but its "number crunching" capabilities make it a powerful tool for data reduction and analysis in general. You can use Excel like a calculator by typing numbers and mathematical operators into a cell and then pressing **Enter**. Most likely, however, you will write your own formulas or choose common functions from Excel's collection and apply them to cell references instead of numbers. Excel is great for doing repetitive calculations quickly.

In this section you will learn how to write some formulas frequently used in biology. Even when you've become proficient at writing formulas in Excel, however, it's still a good idea to ***do a sample calculation by hand (using your calculator) to make sure the formula you entered in Excel gives you the same result.*** If you find a discrepancy, check the formula and check your math. Make sure the answer makes sense.

TABLE A2.1 Operators commonly used for calculations in Excel

Operator	Meaning
+	Addition
–	Subtraction or negation
*	Multiplication
/	Division
%	Percentage
^	Raised to the exponent
:	Range of adjacent cells
,	Multiple, non-adjacent cells

Formulas in Excel always start with an equal sign (=) followed by the cell references, numbers, and operators that make up the formula. Some commonly used operators are shown in Table A2.1. Excel performs the calculations in order from left to right according to the same order of operations used in algebra: first negation (–), then all percentages (%), then all exponentiations (^), then all multiplications and divisions (* or /), and finally all subtractions and additions (– or +). For example, the formula "=100–50/10" would result in "95", because Excel performs division before subtraction. To change the order of operations, enclose part of the formula in parentheses. For example, "=(100–50)/10" would result in "5."

Writing formulas

This section explains how to enter a formula in the correct format and to apply the formula to a range of cells.

Mac

TYPE A FORMULA IN THE ACTIVE CELL

1. Click a cell in which you want the result of the formula to be displayed (the so-called **active cell**). The selected cell will have a green border with a small square in the lower right corner, as in cell N4 in Figure A2.2B. The small square is called the **fill handle** and is used to copy formulas (see "Copying formulas using the fill handle").
2. Type "=" (equal sign).
3. Type the constants, operators, cell references, and functions that you want to use in the calculation. See Table A2.2 for examples.

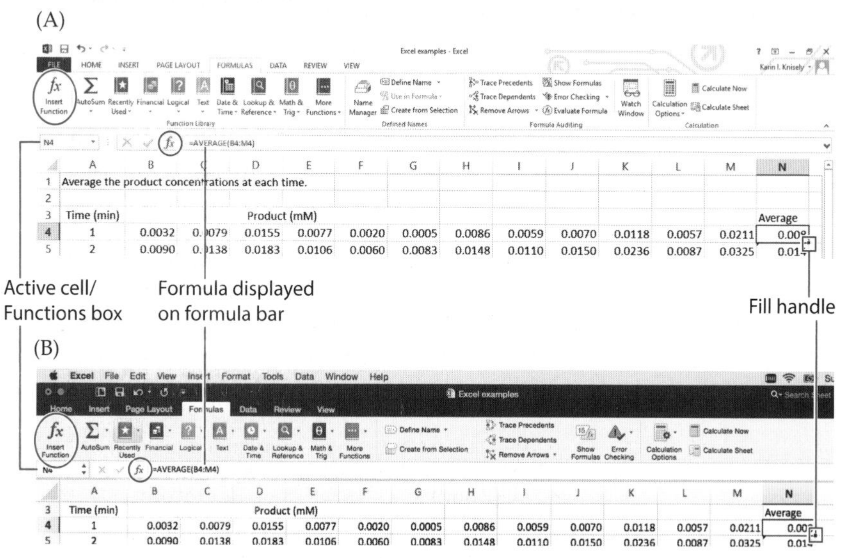

Figure A2.2 The **Formulas** tab has an **Insert Function** button and other commands for doing calculations in (A) Excel 2013, and (B) Excel for Mac 2016.

4. Press **Return**. The result of the calculation is displayed in the active cell. You can view and edit the formula on the **formula bar**.

APPLY A FUNCTION TO VALUES IN ADJACENT CELLS

1. Click a cell in which you want the result of the formula to be displayed.
2. Type "=", the name of the function (e.g., AVERAGE) and then "(".
3. Hold down the left mouse button and select the cells you want to average. If there are a lot of cells to average, click the first cell, hold down the **Shift** key, and then click the last cell in the range. Regardless of the selection method, Excel automatically inserts the first cell and last cell of the range, separated by a colon, in the formula bar (see Figure A2.2B).
4. Type ")" and press **Return**.

APPLY A FUNCTION TO VALUES NOT IN ADJACENT CELLS

1. Click a cell in which you want the result of the formula to be displayed.
2. Type "=", the name of the function (e.g., AVERAGE) and then "(".

TABLE A2.2 Examples of formulas written in Excel

Formula	Meaning
=16*31-42	Because multiplication is done before subtraction, the result is (16 x 31) minus 42.
=A1	The value in cell A1 is assigned to the active cell. If the value in A1 changes, the value will change automatically in all cells that reference A1.
=A1+B1	The sum of the values in cells A1 and B1 is assigned to the active cell. Cell references eliminate the risk of introducing typos when retyping values.
=(O4/H4)^(10/(I4–B4))	The operations in parentheses are done first. The value in cell O4 is divided by the value in cell H4. This number is raised to the power of 10 divided by the difference between the values in cells I4 and B4.
=SUM(A1:A26)	The sum of the values in cells A1 through A26 inclusive is assigned to the active cell. To select a range of adjacent cells, use **Shift**.
=SUM(B4,J4,L4,T4)	The sum of the values in cells B4, J4, L4, and T4 is assigned to the active cell. To select a range of nonadjacent cells, use **Ctrl** (Windows) or ⌘ (Mac).
=AVERAGE(B4:U4)	The average of the values in cells B4 through U4 inclusive is assigned to the active cell. To select a range of adjacent cells, use **Shift**.
=AVERAGE(B4,J4,L4,T4)	The average of the values in cells B4, J4, L4, and T4 is assigned to the active cell. To select a range of nonadjacent cells, use **Ctrl** (Windows) or ⌘ (Mac).
=log(B1)	The log of the value in cell B1 is assigned to the active cell.
=ln(B1)	The natural log of the value in cell B1 is assigned to the active cell.

3. Hold down the ⌘ key and click the cells you want to average. Excel automatically inserts the selected cells, separated by commas, in the formula bar.
4. Type ")" and press **Return**.

SELECT AN EXCEL FUNCTION WITH THE INSERT FUNCTION BUTTON

1. Click a cell in which you want the result of the formula to be displayed.
2. Click **Formulas | Insert Function** (see Figure A2.2B). Or, from any tab, click the **f_x (Insert Function)** button to the left of the formula bar.

3. If the function you're looking for is not shown in the **Most Recently Used** category, click **Show All Functions** in the Formula Builder box and search under **All**.
4. When you locate the function you want to use, click **Insert Function** and then, if necessary, on the formula bar, modify the range of cells to which the function will be applied.

SELECT AN EXCEL FUNCTION FROM THE ACTIVE CELL/FUNCTIONS LIST

1. Click a cell in which you want the result of the formula to be displayed.
2. Type "=" in the active cell. The most recently used function is displayed in the **Active Cell/Functions** box to the left of the formula bar (see Figure A2.2B).
3. Click the down arrow to select a function from the drop-down menu. The Formula Builder pane explains the function you've selected.
4. On the formula bar, enter the range of cells to which the function will be applied.

Windows

TYPE A FORMULA IN THE ACTIVE CELL

1. Click a cell in which you want the result of the formula to be displayed (the so-called **active cell**). The selected cell will have a green border with a small square in the lower right corner, as in Figure A2.2A. The small square is called the **fill handle** and is used to copy formulas (see "Copying formulas using the fill handle").
2. Type "=" (equal sign).
3. Type the constants, operators, cell references, and functions that you want to use in the calculation. See Table A2.2 for examples.
4. Press **Enter**. The result of the calculation is displayed in the active cell. You can view and edit the formula on the **formula bar**.

APPLY A FUNCTION TO VALUES IN ADJACENT CELLS

1. Click a cell in which you want the result of the formula to be displayed.
2. Type "=", the name of the function (e.g., AVERAGE) and then "(".
3. Hold down the left mouse button and select the cells you want to average. If there are a lot of cells to average, click the first cell, hold down the **Shift** key, and then click the last cell in the range. Regardless of the selection method, Excel automatically inserts

the first cell and last cell of the range, separated by a colon, on the formula bar (see Figure A2.2A).

4. Type ")" and press **Enter**.

APPLY A FUNCTION TO VALUES NOT IN ADJACENT CELLS

1. Click a cell in which you want the result of the formula to be displayed.
2. Type "=", the name of the function (e.g., AVERAGE) and then "(".
3. Hold down the **Ctrl** key and click the cells you want to average. Excel automatically inserts the selected cells, separated by commas, on the formula bar.
4. Type ")" and press **Enter**.

SELECT AN EXCEL FUNCTION WITH THE INSERT FUNCTION BUTTON

1. Click a cell in which you want the result of the formula to be displayed.
2. Click **Formulas | Function Library | Insert Function** (see Figure A2.2A). Or, from any tab, click the **f_x (Insert Function)** button to the left of the formula bar.
3. If the function you're looking for is not shown in the Recently Used category, click the down arrow to display another category.
4. When you locate the function you want to use, click it in the **Select a function** list box and then click **OK**.
5. In the **Function Arguments** dialog box, Excel tries to guess the range of cells to which you want to apply the function. If the range is incorrect, use the **Shift** key or the **Ctrl** key as explained previously to select the correct range or type the correct cell references on the formula bar.
6. Click **OK** or press **Enter**.

SELECT AN EXCEL FUNCTION FROM THE ACTIVE CELL/FUNCTIONS LIST

1. Click a cell in which you want the result of the formula to be displayed.
2. Type "=" in the active cell. The most recently used function is displayed in the **Active Cell/Functions** box to the left of the formula bar (see Figure A2.2A).
3. Click the down arrow to select a function from the drop-down menu. The **Function Arguments** dialog box appears.
4. Follow steps 5 and 6 above.

Copying formulas using the fill handle (Mac and Windows)

Quite frequently you may want to perform the same calculation on data contained in cells of neighboring rows or columns. Instead of retyping the formula, you can simply drag the formula into adjacent cells. To do so, follow these steps:

1. Click the cell containing the formula you wish to copy.
2. Locate the fill handle, the small square in the lower right corner of the cell (see Figure A2.2, p. 216).
3. Move the mouse over the fill handle to display cross hairs (+).
4. Hold down the left mouse button and drag the fill handle over the cells you want to "fill" with the formula.
5. Click one of the "filled" cells to make sure the formula was copied correctly. If necessary, edit the formula on the formula bar.

Copying cell values, but not the formula (Mac and Windows)

When you select **Copy**, Excel copies everything in the cell—the formula, the number, and the text. When you select **Paste**, however, you may get a "#REF!" error instead of the entry you expected. This error typically occurs when the connection between the cell references and the formula is lost. To paste the value without the formula, click **Home | Clipboard** (Windows only) | **Paste ▼ | Paste Values**.

Formatting the Spreadsheet (Mac and Windows)

Wrap text

When you type a long text in a cell, it runs into the adjacent cell to the right. If there is text or a numerical entry in the adjacent cell, the long text is hidden (it appears to be cut off). To make the long text visible, you could widen the columns, but then fewer columns will be visible on the screen. To be able to maintain column width and still view the entire text, click the desired cell and then **Home | Alignment** (Windows only) | **Wrap Text**. Excel expands the row height to accommodate the contents of the cell.

Increase or decrease decimal

When a calculation results in a value that has more decimal places than the measurements from which it originated, round up the final value. In other words, make sure that your final answer does not have more decimal places than those in the original measurement.

EXAMPLE: A digital spectrophotometer can read absorbance to the thousandths place, e.g. 0.235. Let's say that after Excel averages 10 replicate absorbance measurements, the result is 0.24158. The average absorbance should be reported as 0.242.

Only round up the *final* calculation result, not the intermediate values of a multi-step calculation, otherwise you will introduce error. To round up the final result, click the desired cell and then **Home | Number** (Windows only) | **Decrease Decimal** repeatedly until the appropriate number of decimal places is displayed.

Format cells

Home | Cells (Windows only) | **Format | Format Cells** opens a dialog box with 6 tabs: Number, Alignment, Font, Border, Fill, and Protection. For most cells containing numbers, **Number | General** works well. To limit the number of decimal places in the selected cells, click **Number | Number**, enter the desired number of decimal places, and then click **OK**. **Number | Scientific** is not recommended, because Excel's use of **E** to represent *exponent* is not accepted by the Council of Science Editors.

Sort data

If you'd like to alphabetize a list of names or arrange numerical values in ascending or descending order, use the **Home | Editing** (Windows only) | **Sort & Filter** command. First, select ***all of the columns that contain data*** that you want to sort. This is important, because if you select only one column, then only the data in that column will be sorted. Data in the adjacent columns will then no longer correspond to the correct data in the sorted column. After selecting all of the data, click **Custom Sort** on the **Sort & Filter** button's drop-down list. Select the criteria according to which you want to sort: by column, by values, and in which order. Then click **OK**.

Tables

Excel worksheets are technically tables that have more than a million rows and more than 16,000 columns. Most of the time you will analyze data and make graphs in Excel without actually printing out the worksheet. However, your instructor may ask you to attach a table containing the raw data in an appendix. This is different from inserting a table in the lab report itself. It is *not necessary* to include a table when you already have a graph that shows the same data. Make *either* a table or a graph—not both—

depending on your objectives (see "Do the Results Section Next" in Chapter 4).

By convention, tables in scientific papers do not have vertical lines to separate the columns, and horizontal lines are used only to separate the table caption from the column headings, the headings from the data, and the data from any footnotes. To format a table like this in Excel, follow these instructions.

Mac

1. If you would like to insert a row for column headings above the data, click a row and then **Home | Insert ▼ | Insert Sheet Rows**. Then type the column headings.
2. Select the cells containing the column headings.
3. Click **Home | Format ▼ | Format Cells** to open the Format Cells dialog box.
4. Click the top and bottom border options on the **Border** tab. Click **OK** to close the box.
5. Now select the cells containing the last row of the table.
6. Click **Home | Format ▼ | Format Cells** to open the Format Cells dialog box.
7. Click the bottom border option on the **Border** tab. Click **OK** to close the box.
8. Select all of the cells in the table, and copy and paste them into your Word document.
9. Type a table caption *above* the table.

Windows

1. If you would like to insert a row for column headings above the data, click a row and then **Home | Cells | Insert ▼ | Insert Sheet Rows**. Then type the column headings.
2. Select the cells containing the column headings.
3. Click **Home | Cells | Format ▼ | Format Cells** to open the Format Cells dialog box.
4. Click the top and bottom border options on the **Border** tab. Click **OK** to close the box.
5. Now select the cells containing the last row of the table.
6. Click **Home | Cells | Format ▼ | Format Cells** to open the Format Cells dialog box.

7. Click the bottom border option on the **Border** tab. Click **OK** to close the box.
8. Select all of the cells in the table, and copy and paste them into your Word document.
9. Type a table caption *above* the table.

The Page Layout tab

The Page Layout tab is where you format your spreadsheet after you've entered data. The buttons on this tab allow you to:

- Change the margins
- Add headers, footers, or page numbers
- Scale the content to fit on one sheet of paper
- Adjust page breaks
- Set the print area
- Repeat row or column titles on multiple-page worksheets

Views

The View tab in Excel gives you options for displaying your workbook(s) on the screen. There are three workbook views that can also be accessed from the status bar in Windows. Page Break Preview is not available in the Mac OS (see Figure A2.1).

- **Normal** view is the default view and is typically used when entering data.
- **Page Layout** view displays margins; headers, footers, and page numbers; and page breaks.
- **Page Break Preview** is handy for adjusting page breaks to keep blocks of text or data together.

Two other views, called **Freeze Panes** and **Split**, are useful to view row and column titles while scrolling through the spreadsheet. **Split** view splits the worksheet into 2 or 4 sections (based on the cell that is selected), each of which can be scrolled independently. Use **Split** view to view different sections of the worksheet simultaneously. The **Freeze Panes** command allows you to lock row and column titles while scrolling through the rest of the worksheet. To lock the titles, click the cell just below the row and just to the right of the column that contains the titles. Then click one of the **Freeze** options.

Adding worksheets to the same file

Each Excel workbook comes with one worksheet, named Sheet1 (see Figure A2.1). You can give this sheet a different name by double-clicking the tab in the lower left corner of the screen. You can also add more worksheets by clicking the + **New sheet** button in Windows or the corresponding + **Insert Sheet** button in the Mac OS. Some good reasons for keeping multiple worksheets in one workbook are:

- To collect replicate data, yet keep individual trials separate.
- To collect data from multiple lab sections for the same experiment, to pool or keep separate as needed.
- To simplify file organization.

You can rearrange the worksheets by dragging a tab left or right. To the left of the sheet tabs, you'll see two sheet tab scroll buttons. The scroll buttons are only available if the workbook contains so many worksheets that their tabs cannot all be displayed at once.

Plotting XY Graphs (Scatter Charts)

Scientists call XY graphs "line graphs," but you should not confuse line graphs with Excel's "line charts." **Line charts** are a special kind of XY graph used to show trends over regular time intervals. Line charts do not space data proportionally on the *x*-axis. For example, intervals of 5, 20, and 50 units would be spaced equally, when in fact there should be 5 units in the first interval, 15 in the second, and 30 in the third. The bottom line is: *When you want to make an XY graph in Excel, choose "scatter charts."*

XY graphs are used to display a relationship between two or more quantitative variables. Before you enter data in an Excel worksheet, you must determine which variable to plot on the *x*-axis and which one on the *y*-axis. By convention, the *x*-axis of the graph shows the independent variable, the one that was manipulated during the experiment. The *y*-axis of the graph shows the dependent variable, the variable that changes in response to changes in the independent variable.

Mac

Enter the data

One data set. In the first row, enter a short, informative title for each column (Figure A2.3). Enter the values for the independent variable in column A. The independent variable is plotted on the *x*-axis. Enter the corresponding values for the dependent variable in column B. The dependent variable responds to the independent variable and is plotted on the *y*-axis.

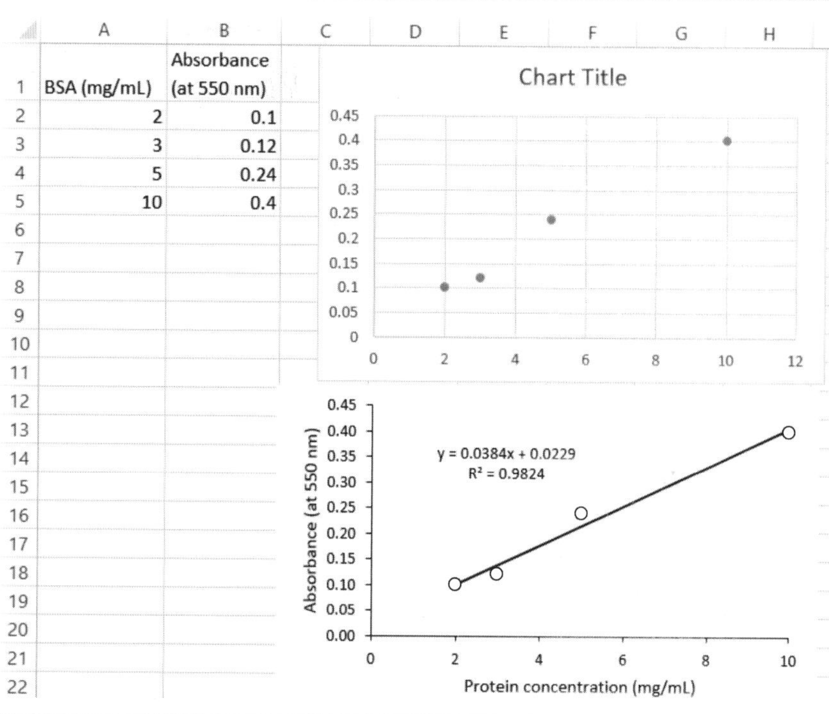

Figure A2.3 Before- and after-formatting a protein standard curve. The data for the independent and dependent variables are entered in adjacent columns. In this example, *x*-axis data (independent variable) are entered in Column A and *y*-axis data (dependent variable) in Column B. The difference in contrast between Excel's default color for the axes and characters—gray—and true black is especially noticeable when the graph is printed in black and white.

More than one data set. If you want to plot more than one data set (line) on the same graph, follow the instructions for entering data for one data set, and then enter the values for the other treatment groups in columns C, D, E, and so on. When plotting multiple data sets on the same graph, it is especially important to enter informative titles for each column, as Excel uses these titles to make a legend. Make the titles descriptive enough to allow the conditions to be distinguished unambiguously, but also keep the titles short. Using Figure A2.4 as an example, notice the difference between a concise title for the legend and one that contains too much information.

FAULTY: Absorbance at 420 nm for WT, Gly

CONCISE: WT, Gly

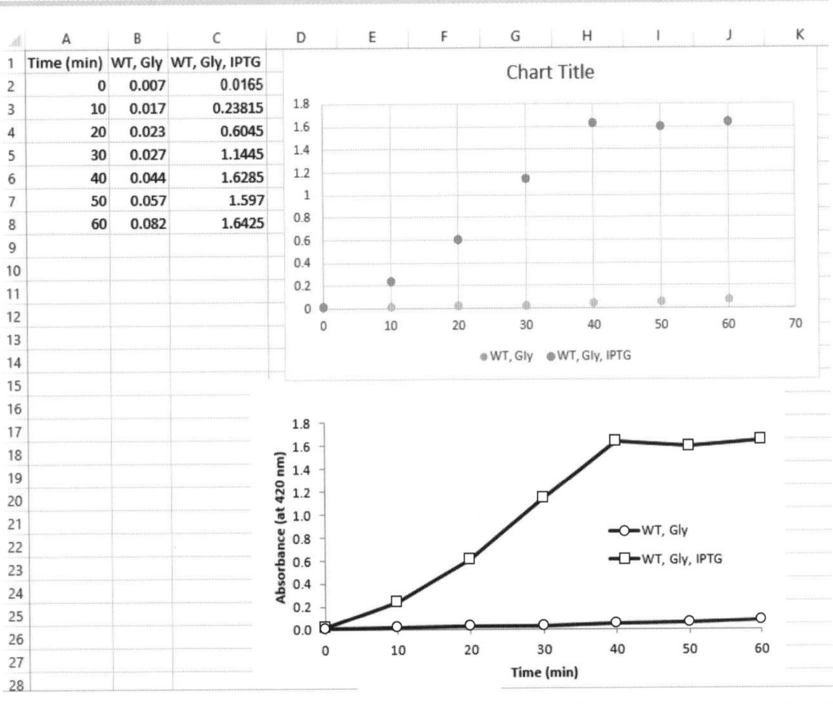

Time (min)	WT, Gly	WT, Gly, IPTG
0	0.007	0.0165
10	0.017	0.23815
20	0.023	0.6045
30	0.027	1.1445
40	0.044	1.6285
50	0.057	1.597
60	0.082	1.6425

Figure A2.4 Before- and after-formatting an XY graph with more than one line. *x*-axis data are entered in Column A and the dependent variable data for each treatment group are entered in the adjacent columns to the right. Excel's default markers are different colored filled circles, which are nearly impossible to distinguish when the graphs are printed in black and white. On the other hand, the CSE symbols in the after formatting graph can be distinguished easily. The legend entries provide the minimum amount of information needed to identify the treatment groups.

Similarly, for Figure 4.7 on p. 72, it is not necessary to include the name of the dependent variable in the legend, because the reader can get this information by looking at the *y*-axis label.

FAULTY: Height in cm for light-grown plants

CONCISE: Light-grown

Select the data

Hold down the left mouse button and select the data.

One data set. It is not necessary to select the column titles. Just select the numerical values.

More than one data set. Select the column titles along with the numerical values so that Excel can make an informative legend. If there are 6 or more data sets, delete the title from Column A (the x-axis variable), otherwise Excel will not plot the lines correctly.

Plot the data

For the time being, click **Insert | X Y (Scatter) | Scatter** to plot the data without any lines. *Scatter* (as opposed to *Line*) plots the data in the correct intervals. When there are measured data, the **markers** (Excel's term for the data points) *must* be displayed. Later, in the "Choose a line" section, we will determine what kind of line (if any) is appropriate for the data based on the purpose of the graph.

Add chart elements

The graph that Excel creates needs to be formatted according to Council of Science Editors' standards. Click the graph to activate the Chart Design tab on the Ribbon. Click the **Add Chart Element** button all the way to the left.

1. Axes: **Primary Horizontal** and **Primary Vertical** are checked by default.
2. Axis Titles: Click **Primary Horizontal** and enter a title for the x-axis. Type the variable and then the units in parentheses. Click **Primary Vertical** and enter a title for the y-axis. The default font color is gray. To make the font black for best contrast on a white background, triple-click the axis title to open the Format Axis Title pane on the right. Click **Text Options | Text Fill | Color: Black**.
3. Chart Title: Change to **None** when the figure will be inserted into a scientific paper. Select **Chart Title | Above Chart** when the figure will be used in a PowerPoint presentation.
4. Data Labels: **None** is the default.
5. Error Bars: **None** is the default.
6. Gridlines: Remove both horizontal and vertical gridlines by unchecking **Primary Major Horizontal** and **Primary Major Vertical**.
7. Legend: **None** when there is only one data set. When there are multiple data sets, click **More Legend Options**. Select **Legend Position | Right** and uncheck **Show the legend without overlapping the chart**. On the graph, drag the legend completely inside the axes. If there is not enough space, identify the data sets in the figure caption.

Format the axes

1. Notice that neither axis has tick marks. To insert tick marks, click the axis to display the Format Axis pane to the right. Click the **bar graph symbol** on the Axis Options tab, and scroll down to **Tick Marks**. Next to **Major type**, choose **Outside**. Leave **Minor type** as **None**.
2. Now click the **paint bucket symbol** and change the **Line** color (*not* the **Fill** color) of the axis from gray to **black** for best contrast.
3. Repeat the process for the other axis. Click the axis and **Axis Options** | the **bar graph symbol | Tick Marks | Major type: Outside**. Then click the **paint bucket symbol | Line | Color: Black**.
4. Check that all the numbers on the *y*-axis have the same number of decimal places. If they don't, click the *y*-axis to open the Format Axis pane to the right. Click **Axis Options | the bar graph symbol | Number**. Change the **Category** from **General** to **Number**, and change the **Decimal places** to **1** or another value that makes sense for the scale of the graph.
5. The data should fill the plot area so that there is little empty space. If necessary, change the limits of the axes by clicking the **bar graph symbol** in the **Format Axis | Axis Options** pane. Then enter the appropriate numbers in **Axis Options | Bounds | Minimum** or **Maximum**. The tick mark intervals can be adjusted under **Units | Major**.

Format the markers

Now let's change the markers from filled colored circles to symbols recommended by the Council of Science Editors for ease of recognition in black and white publications. The CSE hierarchy of symbols is unfilled black circle, filled black circle, unfilled black triangle, filled black triangle, unfilled black square, filled black square (Table A2.3).

1. Click any of the markers to display the Format Data Series pane to the right.
2. Click the **paint bucket symbol | Marker | Fill | Color: White**. This selection makes an unfilled shape. Select **Solid fill | Color: Black** to make a filled shape.
3. In the same pane, click **Border | Solid line | Color: Black**. This selection makes the outline of the symbol black for best contrast on a white background.
4. Finally, to change the size and shape of the marker itself, click the **paint bucket symbol | Marker | Marker Options | Built-in**. Select the type (circle, triangle, or square) and increase the size to at least **7**.

TABLE A2.3 Comparison of Excel and CSE Manual symbol hierarchy (Peterson 1999) for XY graphs

Excel	CSE Manual
Light blue filled circle	Black open circle
Red filled circle	Black filled circle
Gray filled circle	Black open triangle
Yellow filled circle	Black filled triangle
Dark blue filled circle	Black open square
Green filled circle	Black filled square

Choose a line

What kind of line (if any) should you use to best show the relationship between the variables? That depends on the purpose of the graph and your instructor's instructions. Reasons to add a line include

- To show a trend.
- To show a cause and effect relationship between the independent and dependent variables.

If you decide to add a line, should that line be straight or smoothed, or is a mathematical trendline the best option?

- Adding **straight** lines between the markers implies that you are making no assumptions about the results for the conditions between the markers, because you didn't actually measure them. To insert straight lines to connect the markers, click the chart area (the part of the graph outside the axes), and select **Chart Design | Change Chart Type | X Y (Scatter) | Scatter with Straight Lines and Markers**.
- On the other hand, a **smoothed** line may be more appropriate to show differences in response, which reflect the natural variability in a large sample size, rather than strictly the effect of the independent variable. For example, when measuring plant height over time, it would be unlikely for the height of an individual, healthy plant to decrease with time. However, the *average* height of many randomly chosen plants might show a decrease over time. In that situation, inserting a smoothed line on the data points would depict the effect of the independent variable more realistically. To insert smooth lines to connect the markers, click the chart area (the part of the graph outside the axes), and select **Chart Design | Change Chart Type | X Y (Scatter) | Scatter with Smooth Lines and Markers**.

- Finally, when there is an expected mathematical relationship between the variables or you need to predict one variable from the other, insert a **trendline**. For example, from Beer's Law we expect absorbance to be proportional to concentration. Thus a **linear** trendline should be used for these data. On the other hand, enzyme kinetics has established that enzymatic activity increases logarithmically as a function of substrate concentration. Thus a **logarithmic** trendline should be inserted on these data points.

To insert a trendline, click a marker and under **Chart Design | Add Chart Element**, select **Trendline | More Trendline Options**.

1. On the Format Trendline pane to the right, click **the bar graph symbol**. You'll see that **Linear** is the default, but other options are available.
2. If you are using this graph to predict one variable from the other variable, which is usually measured, display the equation of the line. In addition, to show how well (or not) the equation fits the experimental data, display the R-squared value. To do so, check the **Display Equation on chart** and **Display R-squared value on chart** boxes at the bottom of the pane.
3. On the graph itself, drag the text box so that it doesn't overlap the trendline. If necessary, make the text color black.

Format the line

Solid black lines are preferred for best contrast on a white background.
Scatter with straight/smooth lines and markers. To change Excel's default colored lines to solid black lines, click the line to display the Format Data Series pane on the right. Click the **paint bucket symbol | Line** and change the color to **black**.
Trendlines. Click the trendline to display the Format Trendline pane. Click the **paint bucket symbol** and change the color to **black**. Under **Dash type**, choose the first option, the **solid line**. When there are multiple lines on the same graph, different dash types may be used to help distinguish the data sets.

Remove chart border

Figures in scientific papers do not have a border. To remove the border around the graph, click the chart area (*not* the **plot area**, which is the area inside the axes). In the Format Chart Area pane to the right, click **Chart Options | the paint bucket symbol**, and under **Border**, click **No line**.

Import graph into Word document

Compare the before- and after-formatting graphs in Figures A2.3 and A2.4. The gridlines in the before-graph make the graph look cluttered and obscure the trend shown by the data points. The blue markers (which look gray when the graph is printed in a black and white publication) do not stand out against a white background. The crisp format of the after-graph focuses attention on the results.

To copy and paste the graph you made in Excel into your Word document, follow these steps.

1. Click the chart area (not the plot area).
2. Click **Home | Copy** or ⌘+**C**.
3. In your Word document, position the cursor below the paragraph in which you first describe the graph (typically in the Results section). Click **Home | Paste** or ⌘+**V**.
4. Press **Return** after inserting the figure, and then type a figure caption *below* the figure.

Windows

Enter the data

One data set. In the first row, enter a short, informative title for each column (see Figure A2.3). Enter the values for the independent variable in column A. The independent variable is plotted on the x-axis. Enter the corresponding values for the dependent variable in column B. The dependent variable responds to the independent variable and is plotted on the y-axis.
More than one data set. If you want to plot more than one data set (line) on the same graph, follow the instructions for entering data for one data set, and then enter the values for the other treatment groups in columns C, D, E, and so on. When plotting multiple data sets on the same graph, it is especially important to enter informative titles for each column, as Excel uses these titles to make a legend. Make the titles descriptive enough to allow the conditions to be distinguished unambiguously, but also keep the titles short. Using Figure A2.4 as an example, notice the difference between a concise title for the legend and one that contains too much information.

FAULTY: Absorbance at 420 nm for WT, Gly

CONCISE: WT, Gly

Similarly, for Figure 4.7 on p. 72, it is not necessary to include the name of the dependent variable in the legend, because the reader can get this information by looking at the y-axis label.

FAULTY: Height in cm for light-grown plants

CONCISE: Light-grown

Select the data

Hold down the left mouse button and select the data.

One data set. It is not necessary to select the column titles. Just select the numerical values.

More than one data set. Select the column titles along with the numerical values so that Excel can make an informative legend. If there are 6 or more data sets, delete the title from Column A (the *x*-axis variable), otherwise Excel will not plot the lines correctly.

Plot the data

For the time being, click **Insert | Insert Scatter (X, Y) or Bubble Chart ▼ | Scatter** to plot the data without any lines. *Scatter* (as opposed to *Line*) plots the data in the correct intervals. When there are measured data, the **markers** (Excel's term for the data points) *must* be displayed. Later, in the "Choose a line" section, we will determine what kind of line (if any) is appropriate for the data based on the purpose of the graph.

Add chart elements

The graph that Excel creates needs to be formatted according to Council of Science Editors' standards. Click the graph to activate the 3 buttons at the top right of the graph. Click the + **Chart Elements** button to display a checklist of formatting elements. The same list can be accessed by clicking **Chart Tools Design | Add Chart Element**.

1. Axes: **Primary Horizontal** and **Primary Vertical** are checked by default.
2. Axis Titles: Click **Axis Titles**. Enter a title for the *y*-axis. Type the variable and then the units in parentheses. In the same way, enter a title for the *x*-axis. Triple-click each title and change the font color from gray to **black** on the mini toolbar.
3. Chart Title: Change to **None** when the figure will be inserted into a scientific paper. Select **Chart Title | Above Chart** when the figure will be used in a PowerPoint presentation.
4. Data Labels: **None** is the default.
5. Error Bars: **None** is the default.
6. Gridlines: Uncheck **Gridlines**.
7. Legend: **None** when there is only one data set. When there are multiple data sets, click **More Options**. Select **Legend Position | Right** on the Format Legend pane to the right, and uncheck **Show**

the legend without overlapping the chart. On the graph, drag the legend completely inside the axes. If there is not enough space, identify the data sets in the figure caption.

Format the axes

1. Notice that neither axis has tick marks. To insert tick marks, click the axis to display the Format Axis pane to the right. Click the **bar graph symbol** on the Axis Options tab and scroll down to **Tick Marks**. Next to **Major type**, choose **Outside**. Leave **Minor type** as **None**.
2. Now click the **paint bucket symbol** and change the **Line** color (*not* the **Fill** color) of the axis from gray to **black**.
3. Repeat the process for the other axis. Click the axis, and click **Axis Options | the bar graph symbol | Tick Marks | Major type: Outside**. Then click the **paint bucket symbol | Line | Color: Black**.
4. Check that all the numbers on the *y*-axis have the same number of decimal places. If they don't, click the *y*-axis to open the Format Axis pane. Click **Axis Options | the bar graph symbol | Number**. Change the **Category** from **General** to **Number**, and change the **Decimal places** to **1** or another value that makes sense for the scale of the graph.
5. The data should fill the plot area so that there is little empty space. If necessary, change the limits of the axes by clicking the **bar graph symbol** in the **Format Axis | Axis Options** pane. Then enter the appropriate numbers in **Axis Options | Bounds | Minimum** or **Maximum**. The tick mark intervals can be adjusted under **Units | Major**.

Format the markers

Now let's change the markers from filled colored circles to symbols recommended by the Council of Science Editors for ease of recognition in black and white publications. The CSE hierarchy of symbols is unfilled black circle, filled black circle, unfilled black triangle, filled black triangle, unfilled black square, filled black square (Table A2.3, p. 229).

1. Click any of the markers to display the Format Data Series pane to the right.
2. Click the **paint bucket symbol | Marker | Fill | Color: White**. This selection makes an unfilled shape. Select **Solid fill | Color: Black** to make a filled shape.

3. In the same pane, click **Border | Solid line | Color: Black**. This selection makes the outline of the symbol black for best contrast on a white background.
4. Finally, to change the size and shape of the marker itself, click the **paint bucket symbol | Marker | Marker Options | Built-in**. Select the type (circle, triangle, or square) and increase the size to at least **7**.

Choose a line

What kind of line (if any) should you use to best show the relationship between the variables? That depends on the purpose of the graph and your instructor's instructions. Reasons to add a line include

- To show a trend.
- To show a cause and effect relationship between the independent and dependent variables.

If you decide to add a line, should that line be straight or smoothed, or is a mathematical trendline the best option?

- Adding **straight** lines between the markers implies that you are making no assumptions about the results for the conditions between the markers, because you didn't actually measure them. To insert straight lines to connect the markers, click the **chart area** (the part of the graph outside the axes), and select **Chart Tools Design | Change Chart Type | X Y (Scatter) | Scatter with Straight Lines and Markers**.
- On the other hand, a **smoothed** line may be more appropriate to show differences in response, which reflect the natural variability in a large sample size, rather than strictly the effect of the independent variable. For example, when measuring plant height over time, it would be unlikely for the height of an individual, healthy plant to decrease with time. However, the *average* height of many randomly chosen plants might show a decrease over time. In that situation, inserting a smoothed line on the data points would depict the effect of the independent variable more realistically. To insert smooth lines to connect the markers, click the **chart area** (the part of the graph outside the axes), and select **Chart Tools Design | Change Chart Type | X Y (Scatter) | Scatter with Smooth Lines and Markers**.
- Finally, when there is an expected mathematical relationship between the variables or you need to predict one variable from the other, insert a **trendline**. For example, from Beer's Law we expect absorbance to be proportional to concentration. Thus a **linear** trendline should be used for these data. On the other hand,

enzyme kinetics has established that enzymatic activity increases logarithmically as a function of substrate concentration. Thus a **logarithmic** trendline should be inserted on these data points.

To insert a trendline, click a marker and under **Chart Tools Design | Add Chart Element**, select **Trendline | More Trendline Options**.

1. On the Format Trendline pane to the right, click the **bar graph symbol**. You'll see that **Linear** is the default, but other options are available.
2. If you are using this graph to predict one variable from the other variable, which is usually measured, display the equation of the line. In addition, to show how well (or not) the equation fits the experimental data, display the R-squared value. To do so, check the **Display Equation on chart** and **Display R-squared value on chart** boxes at the bottom of the pane.
3. On the graph itself, drag the text box so that it doesn't overlap the trendline. If necessary, make the text color black.

Format the line

Solid black lines are preferred for best contrast on a white background.
Scatter with straight/smooth lines and markers. To change Excel's default colored lines to solid black lines, click the line to display the Format Data Series pane on the right. Click the **paint bucket symbol | Line** and change the color to **black**.
Trendlines. Click the trendline to display the Format Trendline pane. Click the **paint bucket symbol** and change the color to **black**. Under **Dash type**, choose the first option, the **solid line**. When there are multiple lines on the same graph, different dash types may be used to help distinguish the data sets.

Remove chart border

Figures in scientific papers do not have a border. To remove the border around the graph, click the chart area (*not* the **plot area**, which is the area inside the axes). In the Format Chart Area pane to the right, click **Chart Options | the paint bucket symbol**, and under **Border**, click **No line**.

Import graph into Word document

Compare the before- and after-formatting graphs in Figures A2.3 and A2.4. The gridlines in the before-graph make the graph look cluttered and obscure the trend shown by the data points. The colored markers (which look gray when the graph is printed in a black and white publication) do not stand out against a white background. The crisp format of the after-graph focuses attention on the results.

To copy and paste the graph you made in Excel into your Word document, follow these steps.

1. Click the chart area (not the plot area).
2. Click **Home | Copy** or **Ctrl+C**.
3. In your Word document, position the cursor below the paragraph in which you first describe the graph (typically in the Results section). Click **Home | Paste** or **Ctrl+V**.
4. Press **Enter** after inserting the figure, and then type a figure caption *below* the figure.

Saving and Applying Chart Templates

It takes a fair amount of time to format graphs in Excel because there are so many options. Fortunately, you can save the format as a **chart template** to save yourself time the next time you have to make the same type of chart.

Mac

1. Click the chart and then **Chart Design | Change Chart Type | Save As Template**. Give the template a specific name, for example, **Scatter with 1 linear trendline**. Use the default location (in the Chart Templates folder).
2. To apply this format to new data, select the data and on the Insert tab on the Ribbon, click the button with the bar graph with the circles underneath (**Insert Area, Stock, Surface or Radar Chart**) and then **Templates**. Select the desired template.
3. For some strange reason, when scatter chart templates are applied to new data, Excel creates a chart with incorrect formatting. To correct the problem, click the chart, and then click **Chart Design | Switch Row/Column** twice.
4. Finally, click each Axis Title text box and type the desired title. If necessary, adjust the length of the axes to eliminate empty space as described in "Format the axes."

Windows

1. Right-click the chart and select **Save As Template** from the drop-down menu. Give the template a specific name, for example, **Scatter with 1 linear trendline**. Use the default **Save as type** (Chart Template Files).

2. To apply this format to new data, select the data and click **Insert | Charts diagonal arrow | All Charts | Templates**. Double-click the saved template that you want to apply.
3. For some strange reason, when scatter chart templates are applied to new data, Excel creates a chart with incorrect formatting. To correct the problem, click the chart, and then click **Chart Tools Design | Switch Row/Column** twice.
4. Finally, click each Axis Title text box and type the desired title. If necessary, adjust the length of the axes to eliminate empty space as described in "Format the axes."

Adding Data after Graph Has Been Formatted (Mac and Windows)

If you have gone to all this trouble to format a graph, and realize after the fact that you need to add another data point, the last thing you want to do is to start from scratch. Fortunately, Excel makes it possible to add data after the graph has been made.

To incorporate additional data points in the same series

Insert a new row in the worksheet for the new data points. To do this, click a cell in the row below where the new row is to be inserted. Then click **Home | Cells** (Windows only) | **Insert ▼ | Insert Sheet Rows**. Enter the value for the *x*-axis in Column A and that for the *y*-axis in Column B. The graph is automatically updated for these values.

To incorporate additional lines on the same graph

When adding data sets to an existing graph, the *x*-axis values remain the same, but additional *y*-axis values that represent data for a different treatment or condition have to be added. The *y*-axis values for the existing line are already in Column B. Type the *y*-axis values for the second line in Column C. For each additional line, enter data in the next column to the right.

1. To add the new data to the existing graph, right-click anywhere in the chart area or the plot area and click **Select Data** (Windows) or **Chart Design | Select Data** (Mac).

   Mac. In the **Select Data Source** dialog box, click the + button below the **Legend Entries (Series)** list box.

   Windows. In the **Select Data Source** dialog box, click the **Add** button in the **Legend Entries (Series)** list box.

2. Enter a descriptive title for each data set, which will be used in the legend.

   Mac. To the right of the **Legend Entries (Series)** list box, click the box next to **Name**. Click the cell reference containing the column heading, or enter a short title for the legend.

   Windows. In the **Edit Series** dialog box, under **Series name**, click the cell reference containing the column heading, or enter a short title for the legend.

3. Enter the range of values for the *x*-axis.

   Mac. Click the box next to **X values**. Then click the icon with the red arrow pointing to a cell in a worksheet. Select the values in Column A. Click the icon again.

   Windows. Under **Series X values**, click the icon with the red arrow pointing to a cell in a worksheet. Select the values in Column A. Click the icon again.

4. Enter the range of values for the *y*-axis for the data to be added.

   Mac. Click the box next to **Y values**. Then click the icon with the red arrow pointing to a cell in a worksheet. Select the values in Column C. Click the icon again.

   Windows. Under **Series Y values**, click the icon with the red arrow pointing to a cell in a worksheet. Select the values in Column C. Click the icon again.

5. Repeat Steps 2–4 for each data set. Then Click **OK** to close the **Select Data Source** dialog box.
6. Make the new symbol(s) and line(s) black for best contrast as described under "Format Markers" and "Format the Line."

To change the legend titles

The easiest way to make descriptive legend titles is to make descriptive column headings for your data. Then select the headings along with the data when you make the graph. However, if you need to change the titles after your graph has been finalized, follow these instructions.

1. Click a marker of the data series whose title you want to change. In the Mac OS, click **Chart Design | Select Data**. In Windows, right-click a marker and click **Select Data** from the drop-down menu.
2. In the Select Data Source dialog box of the Mac OS, select the content next to **Name**, and type a new name for the legend entry. Click **OK** to exit. In Windows, select the data set to be edited and then click **Legend Entries (Series) | Edit**. Under **Series name**, type a new name for the legend entry. Click **OK** twice to exit.

Multiple Lines on an XY Graph

Plotting multiple lines on one graph is often the most efficient way to compare the results from several different treatments. How many lines should you put on one set of axes? The CSE Manual recommends no more than eight, but use common sense. You should be able to follow each line individually, and the graph should not look cluttered.

Follow the instructions in the "Plotting XY Graphs" section, particularly for entering and selecting the data. Compare the before- and after-formatting graphs in Figure A2.4 (p. 226) In the before-graph, Excel's default colored markers are nearly impossible to distinguish when the graphs are printed in black and white. In the after-graph, on the other hand, there is no ambiguity regarding either the identity of the data sets or the trends.

Plotting Bar Graphs

Bar graphs are used to compare individual data sets when one of the variables is categorical (not quantitative). By convention, the feature that all the columns have in common (the variable that was measured) lies on the axis parallel to the columns. Excel distinguishes between two basic types of bar graphs: column charts and bar charts. **Column charts** are bar graphs with vertical bars. **Bar charts** are bar graphs with horizontal bars. Bar charts are more practical than column charts when the category labels are long (compare Figures A2.5A and A2.5B).

Enter the labels for the categories in Column A and the quantitative data in Column B. The categories should be sequential on the graph, with the control treatment column farthest left. When deciding what order to enter the categories in the worksheet, remember that the *lowest* row number contains the category label for the *leftmost* vertical bar or the *lowest* horizontal bar. If there is no particular order to the categories, arranging the bars from shortest to longest (or vice versa) makes the results easier to comprehend.

Mac

Enter and select the data

1. In the first row, enter a short, informative title for each column. Enter the categories in column A. Enter the corresponding values for the quantitative variable in column B.
2. Hold down the left mouse button and select the data (without the column headings). To make a column chart with vertical bars, click **Insert | Column | Clustered Column**. To make a bar chart with horizontal bars, click **Insert | Bar | Clustered Bar**.

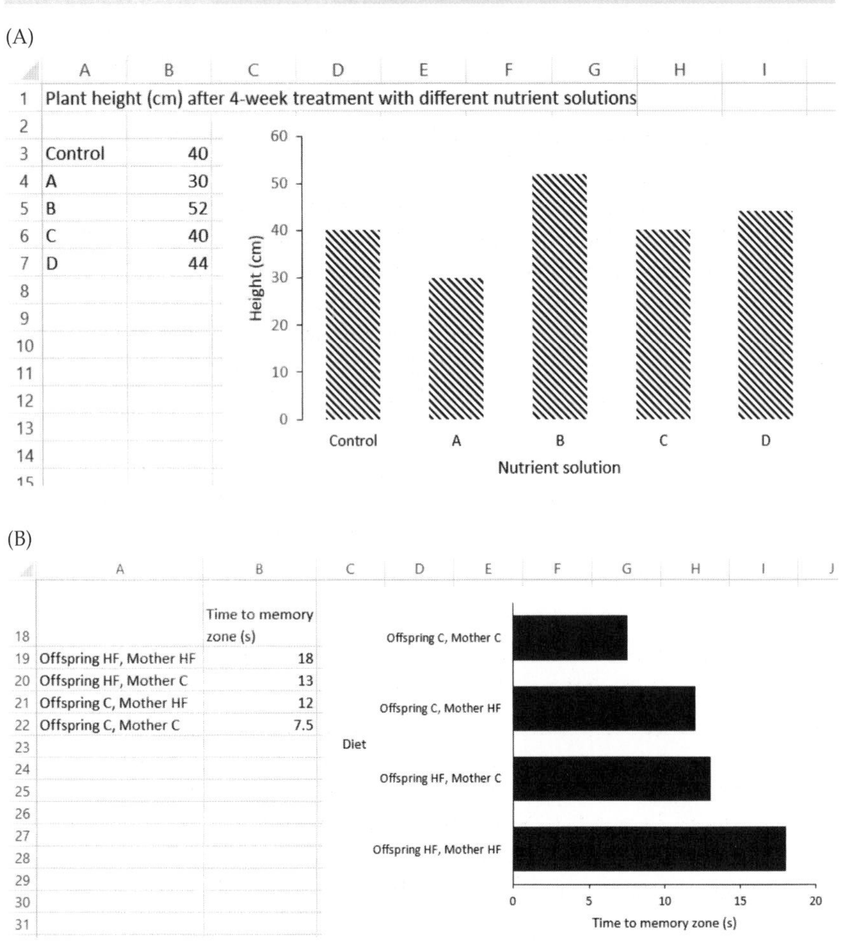

Figure A2.5 Two types of bar graphs. For both types, the labels for the categorical variable are entered in Column A; the data for the response variable are entered in Column B. (A) Column charts have vertical columns. The data for the control are entered in the lowest numbered row to position the corresponding column farthest left on the graph. This column chart has no baseline on the categorical axis. The requirements of the publication (or your instructor) determine whether the line should be present or not. (B) Bar charts have horizontal columns. The category label for the lowest bar (and its data) is entered in the lowest numbered row. Bar charts accommodate long category labels. This bar chart has a baseline on the categorical axis.

Add chart elements

The graph that Excel creates needs to be formatted according to Council of Science Editors' standards. Click the graph to activate the Chart Design tab on the Ribbon. Click the **Add Chart Element** button all the way to the left.

1. Axes: **Primary Horizontal** and **Primary Vertical** are checked by default.
2. Axis Titles: The title for the response axis (the one with the quantitative variable) should have the name of the variable with the units in parentheses. The title for the categorical axis should be a general term that covers the specific categories. The default text font color is gray. To make the font black for best contrast on a white background, triple-click the axis title to open the Format Axis Title pane on the right. Click **Text Options | Text Fill | Color: Black**.
3. Chart Title: Change to **None** when the figure will be inserted into a scientific paper. Select **Chart Title | Above Chart** when the figure will be used in a PowerPoint presentation.
4. Data Labels: **None** is the default.
5. Error Bars: **None** is the default.
6. Gridlines: Remove the gridlines by unchecking **Primary Major Horizontal** and **Primary Major Vertical**.
7. Legend: **None** when there is only one data set. When there are multiple data sets (as in clustered column charts), click **More Legend Options**. Select **Legend Position | Right** and uncheck **Show the legend without overlapping the chart**. On the graph, drag the legend completely inside the axes. If there is not enough space, identify the data sets in the figure caption.

Format the axes

1. Notice that the axis with the quantitative (response) variable does not have tick marks. To insert tick marks, click the axis to display the Format Axis pane to the right. Click the **bar graph symbol**, and scroll down to **Tick Marks**. Next to **Major type**, choose **Outside**. Leave **Minor type** as **None**.
2. Now click the **paint bucket symbol** and change the **Line** color (*not* the **Fill** color) of the axis from gray to **black** for best contrast.
3. According to the CSE Manual, the axis with the categorical variable may or may not be visible, but all the columns must be aligned as if there were a baseline. To remove the line, click the

axis to open the Format Axis pane. Then click **Axis Options | the paint bucket symbol | Line | No line**. Alternatively, make the line black by clicking **Line | Color: Black.** To make the text font black, click **Text Options | Text Fill | Color: Black**.

Adjust bar width

All of the bars in the graph should be the same width, and the bars should always be wider than the space between them. To adjust the width of the bars, click any one of them to open the Format Data Series pane. Click **the bar graph symbol**, and drag the slider for **Gap width** to the left to decrease the gap and simultaneously increase the width of the bars.

Change bar color

Still in the Format Data Series pane, click **the paint bucket symbol | Fill | Solid fill | Color: Black** to make the bars black against a white background for best contrast in black-and-white publications. To make white bars with a black border, click **the paint bucket symbol | Fill | No fill** and then **Border | Solid line | Color: Black**. Alternatively, for good contrast with less ink, click **the paint bucket symbol | Fill | Pattern Fill** and choose one of the heavier line patterns.

Remove chart border

Figures in scientific papers do not have a border. To remove the border around the graph, click the chart area (*not* the **plot area**, which is the area inside the axes). In the Format Chart Area pane to the right, click **Chart Options | the paint bucket symbol**, and under **Border**, click **No line**.

Save as a chart template

To save the bar graph format as a chart template, see "Saving and Applying Chart Templates" on p. 236.

Import graph into Word document

To copy and paste the graph you made in Excel into your Word document, follow these steps.

1. Click the chart area (not the plot area).
2. Click **Home | Copy** or **⌘+C**.
3. In your Word document, position the cursor below the paragraph in which you first describe the graph (typically in the Results section). Click **Home | Paste** or **⌘+V**.
4. Press **Return** after inserting the figure, and then type a figure caption *below* the figure.

Windows

Enter and select the data

1. In the first row, enter a short, informative title for each column. Enter the categories in column A. Enter the corresponding values for the quantitative variable in column B.
2. Hold down the left mouse button and select the data (without the column headings). To make a column chart with vertical bars, click **Insert | Charts | Insert Column Chart | 2-D Column**. To make a bar chart with horizontal bars, click **Insert | Charts | Insert Bar Chart | 2-D Bar**.

Add chart elements

The graph that Excel creates needs to be formatted according to Council of Science Editors' standards. Click the graph to activate the 3 buttons at the top right of the graph. Click the + **Chart Elements** button to display a checklist of formatting elements.

1. Axes: **Primary Horizontal** and **Primary Vertical** are checked by default.
2. Axis Titles: The title for the response axis (the one with the quantitative variable) should have the name of the variable with the units in parentheses. The title for the categorical axis should be a general term that covers the specific categories. The default text font color is gray. Triple-click each title and change the font color from gray to **black** on the mini toolbar.
3. Chart Title: Change to **None** when the figure will be inserted into a scientific paper. Select **Chart Title | Above Chart** when the figure will be used in a PowerPoint presentation.
4. Data Labels: **None** is the default.
5. Error Bars: **None** is the default.
6. Gridlines: Uncheck **Gridlines**.
7. Legend: **None** when there is only one data set. When there are multiple data sets (as in clustered column charts), click **More Options**. In the Format Legend pane, click the **bar graph symbol**. Select **Legend Position | Right** and uncheck **Show the legend without overlapping the chart**. On the graph, drag the legend completely inside the axes. If there is not enough space, identify the data sets in the figure caption.

Format the axes

1. Notice that the axis with the quantitative (response) variable does not have tick marks. To insert tick marks, click the axis to display the Format Axis pane to the right. Click the **bar graph symbol**, and scroll down to **Tick Marks**. Next to **Major type**, choose **Outside**. Leave **Minor type** as **None**.
2. Now click the **paint bucket symbol** and change the **Line** color (*not* the **Fill** color) of the axis from gray to **black** for best contrast.
3. According to the CSE Manual, the axis with the categorical variable may or may not be visible, but all the columns must be aligned as if there were a baseline. To remove the line, click the axis to open the Format Axis pane. Then click **Axis Options | the paint bucket symbol | Line | No line**. Alternatively, make the line black by clicking **Line | Color: Black.** To make the text font black, click **Text Options | Text Fill | Color: Black**.

Adjust bar width

All of the bars in the graph should be the same width, and the bars should always be wider than the space between them. To adjust the width of the bars, click any one of them to open the Format Data Series pane. Click **the bar graph symbol**, and drag the slider for **Gap Width** to the left to decrease the gap and simultaneously increase the width of the bars.

Change bar color

Still in the Format Data Series pane, click **the paint bucket symbol | Fill | Solid fill | Color: Black** to make the bars black against a white background for best contrast in black-and-white publications. To make white bars with a black border, click the **paint bucket symbol | Fill | No fill** and then **Border | Solid line | Color: Black**. Alternatively, for good contrast with less ink, click the **paint bucket symbol | Fill | Pattern Fill** and choose one of the heavier line patterns.

Remove chart border

Figures in scientific papers do not have a border. To remove the border around the graph, click the chart area (*not* the **plot area**, which is the area inside the axes). In the Format Chart Area pane to the right, click **Chart Options | the paint bucket symbol**, and under **Border**, click **No line**.

Save as a chart template

To save the bar graph format as a chart template, see "Saving and Applying Chart Templates" on pp. 236–237.

Import graph into Word document

To copy and paste the graph you made in Excel into your Word document, follow these steps.

1. Click the chart area (not the plot area).
2. Click **Home | Copy** or **Ctrl+C**.
3. In your Word document, position the cursor below the paragraph in which you first describe the graph (typically in the Results section). Click **Home | Paste** or **Ctrl+V**.
4. Press **Enter** after inserting the figure, and then type a figure caption *below* the figure.

Clustered Column Charts

Clustered columns may represent the results of different treatments after the same period of time or the results of the same treatments after different periods of time (Figure A2.6). Each column in the cluster must be easy to distinguish from its neighbor. Colorful columns that look good on your computer screen may turn out to be the same shade of gray when the graph is printed on a black-and-white printer. If the person who evaluates your work receives a black-and-white copy of your paper, be sure to proofread the hardcopy and check that it's clear which treatments the columns represent.

To make a clustered column chart, follow the instructions for plotting bar graphs, with the following differences:

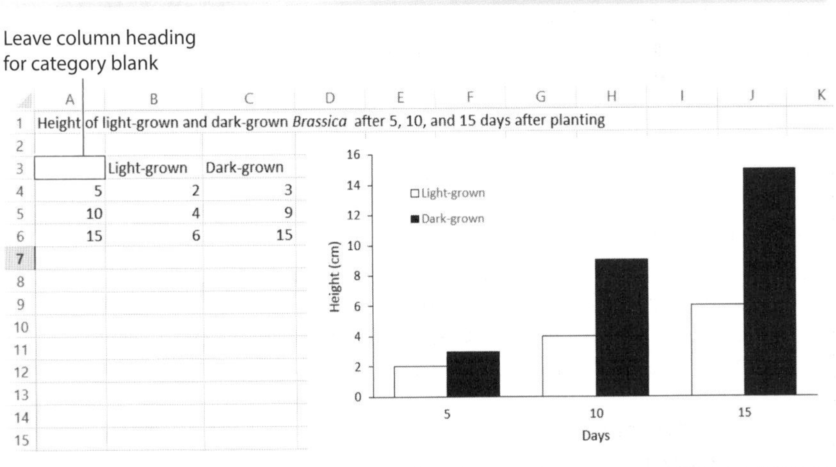

Figure A2.6 Final form of a clustered column chart after formatting in Excel. Additional changes can be made in a Word document after copying and pasting the graph from Excel.

- The response (quantitative) values for the bars in each cluster are entered in Column B, C, and so on.
- When entering column titles for the data, do not enter a title for Column A, otherwise Excel will not format the graph correctly.
- To adjust the width and spacing of the bars, click one of them to open the Format Data Series pane. Click **the bar graph symbol**, and drag the slider for **Gap Width** to the left to increase the width of the columns. In the same pane, adjust **Series Overlap** to **0%** to eliminate the space between the columns in each cluster.

Pie Graphs

Pie graphs are commonly used to show financial data, but they are seldom used in biology research papers. The CSE Manual recommends a table rather than a pie graph for showing percentage of constituents out of the whole. If you would like to make a pie graph, however, follow these instructions.

Mac

Enter and select the data

1. In the first row, enter a short, informative title for each column. These column titles are for your information only; they are not used to make the pie chart. Enter the categories in column A. Enter the corresponding values for the quantitative variable in column B.
2. Hold down the left mouse button and select the data (without the column headings). Click **Insert | Pie | Pie**.

Add chart elements

The graph that Excel creates needs to be formatted according to Council of Science Editors' standards (Figure A2.7). Click the graph to activate the Chart Design tab on the Ribbon. Click the **Add Chart Element** button all the way to the left.

1. Uncheck **Chart Title** unless the figure will be used in a PowerPoint presentation.
2. Uncheck **Legend**.
3. Check **Data Labels | More Data Label Options** to open the Format Data Labels pane on the right. Under **Label Options | Label Contains**, check **Category name** and **Percentage**. Uncheck **Value** and **Show leader lines**. Under **Label Position**, check **Outside end**.

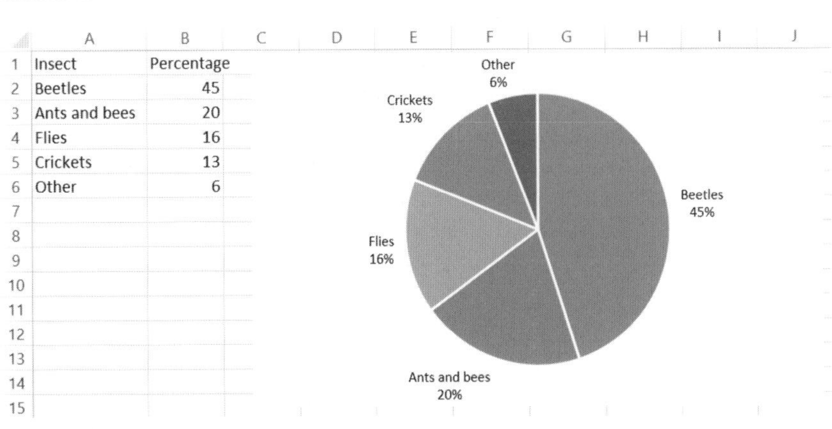

Figure A2.7 Final form of a pie chart after formatting in Excel. Additional changes can be made in a Word document after copying and pasting the graph from Excel.

4. Click a data label and then select **Text Options | Text Fill | Solid fill | Color: Black** on the Format Data Labels pane on the right.

Remove chart border

Figures in scientific papers do not have a border. To remove the border around the graph, click the chart area. In the Format Chart Area pane to the right, click **Chart Options | the paint bucket symbol**, and under **Border**, click **No line**.

Save as a chart template

To save the pie chart format as a chart template, see "Saving and Applying Chart Templates" on p. 236.

Import graph into Word document

To copy and paste the graph you made in Excel into your Word document, follow these steps:

1. Click the chart area.
2. Click **Home | Copy** or **⌘+C**.
3. In your Word document, position the cursor below the paragraph in which you first describe the graph (typically in the Results section). Click **Home | Paste** or **⌘+V**.
4. Press **Return** after inserting the figure, and then type a figure caption *below* the figure.

Windows

Enter and select the data

1. In the first row, enter a short, informative title for each column. These column titles are for your information only; they are not used to make the pie chart. Enter the categories in column A. Enter the corresponding values for the quantitative variable in column B.
2. Hold down the left mouse button and select the data (without the column headings). Click **Insert | Charts | Insert Pie or Doughnut Chart | 2-D Pie**.

Add chart elements

The graph that Excel creates needs to be formatted according to Council of Science Editors' standards. Click the graph to activate the 3 buttons at the top right of the graph. Click the + **Chart Elements** button to display a checklist of formatting elements.

1. Uncheck **Chart Title** unless the figure will be used in a PowerPoint presentation.
2. Uncheck **Legend**.
3. Check **Data Labels | Data Callout**.
4. Right-click a data callout to select all of the callouts. Then select **Outline | No outline** from the mini toolbar to remove the border around each callout.
5. Right-click a data callout again to select all of the callouts. Then select **Font | Font color: Black** from the larger pop-up menu to change the lettering from gray to black for best contrast.

Remove chart border

Figures in scientific papers do not have a border. To remove the border around the graph, click the chart area. In the Format Chart Area pane to the right, click **Chart Options | the paint bucket symbol**, and under **Border**, click **No line**.

Save as a chart template

To save the pie chart format as a chart template, see "Saving and Applying Chart Templates" on pp. 236–237.

Import graph into Word document

To copy and paste the graph you made in Excel into your Word document, follow these steps.

1. Click the chart area.
2. Click **Home | Copy** or **Ctrl+C**.
3. In your Word document, position the cursor below the paragraph in which you first describe the graph (typically in the Results section). Click **Home | Paste** or **Ctrl+V**.
4. Press **Enter** after inserting the figure, and then type a figure caption *below* the figure.

Error Bars and Variability (Mac and Windows)

The reliability of scientific data depends on good experimental design, the skill and experience of the person collecting the data, the reliability of the equipment, and the proven effectiveness of the methods and procedures, among other things. Even when all of these factors have been optimized, there is still a strong likelihood that there will be variability in the measurement data. For example, different students measuring the same sample with the same equipment may come up with different results. Seasoned scientists measuring the same variable in replicate experiments are likely to find slight variations in their data. Genetically identical seeds from the same lot, planted in the same soil and watered at the same time, may germinate at different times and grow at different rates. How can we be confident that our results accurately represent the phenomena we are trying to understand when there is variability in the measurement data?

One way to depict variability is to show all of the measured data on a scatterplot, as in Figure A2.8A. This approach, however, makes it hard to see if there is a trend or relationship between the dependent and independent variables. To reduce the amount of data and begin to make sense of the values, we can take the average of multiple measurements as our best estimate of the true value. In statistics, the average is called the arithmetic mean and it is calculated by dividing the sum of all the values by the number of values. The formula for calculating the mean value in Excel is

=AVERAGE(…),

where "…" is the range of cells to be averaged.

A graph of the mean values is less cluttered (Figure A2.8B), but potentially important information about the variability has been lost. There are two common statistical methods for describing variability: standard deviation and standard error of the mean. Both measures are based on a statistic called variance, which describes how far each measurement value is from the mean. Standard deviation is the square root of the variance and standard error is the standard deviation divided by the square root of the number of measurements. Your handheld calculator and Excel both make it easy to calculate standard deviation and standard error. It is worth your

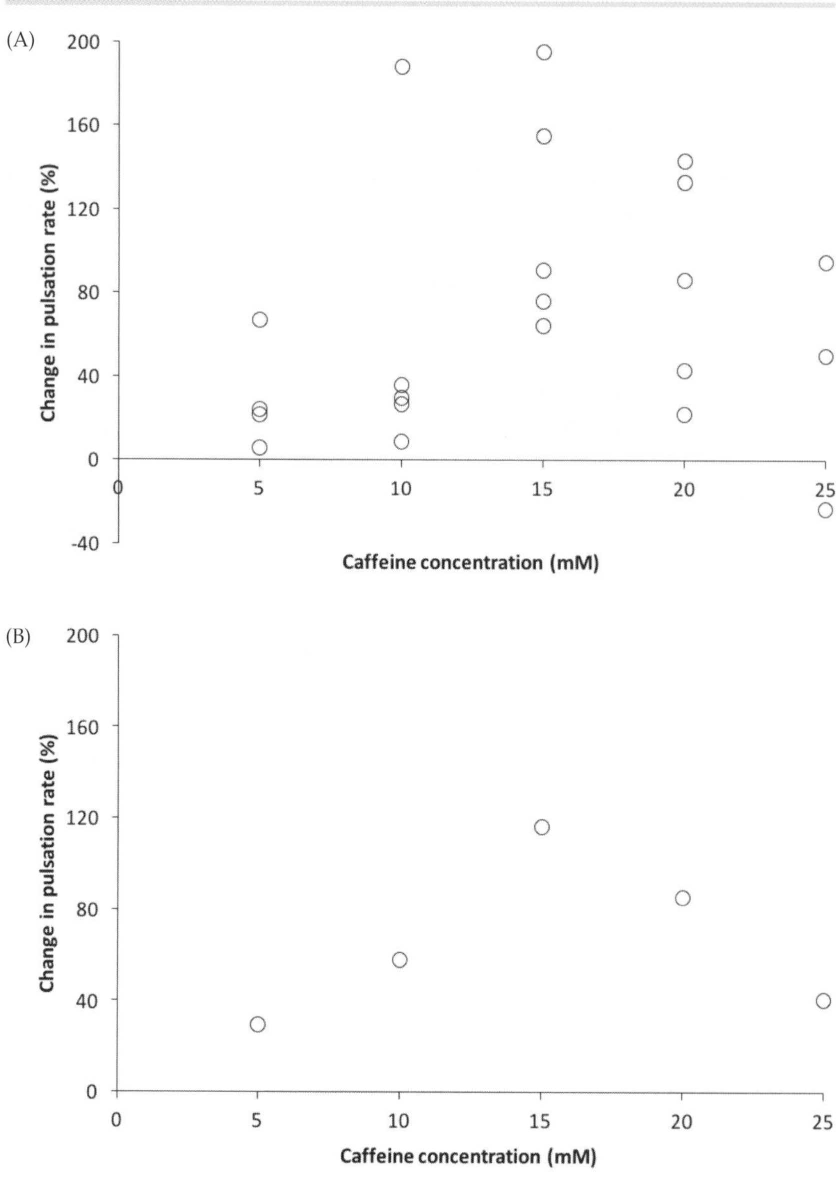

Figure A2.8 (A) All measured data plotted as a scatter graph, and (B) Mean values for effect of caffeine on change in pulsation rate of blackworms. Mean values reduce the amount of data but they do not reflect the degree of variability in the raw values.

while, however, to take a statistics course to learn how these formulas are derived and how to use and interpret statistics appropriately.

Adding error bars about the means

Standard deviation and standard error can be depicted graphically in the form of error bars around the means. To add error bars to the mean values, follow these steps:

1. Select an empty cell to enter the formula for standard deviation, which is

   =STDEV(…)

   where "…" is the range of cells that was averaged to calculate the mean. As shown in Figure A2.9, the four rates for the 5 mM caffeine solution were averaged using the formula "=AVERAGE(B3:B6)" and the mean value is given in cell B28.

2. Select another empty cell to enter the formula for standard error, which is

   $\text{STDEV}/\sqrt{\text{N}}$

   or, in Excel format, =Cell reference for STDEV/SQRT(N)

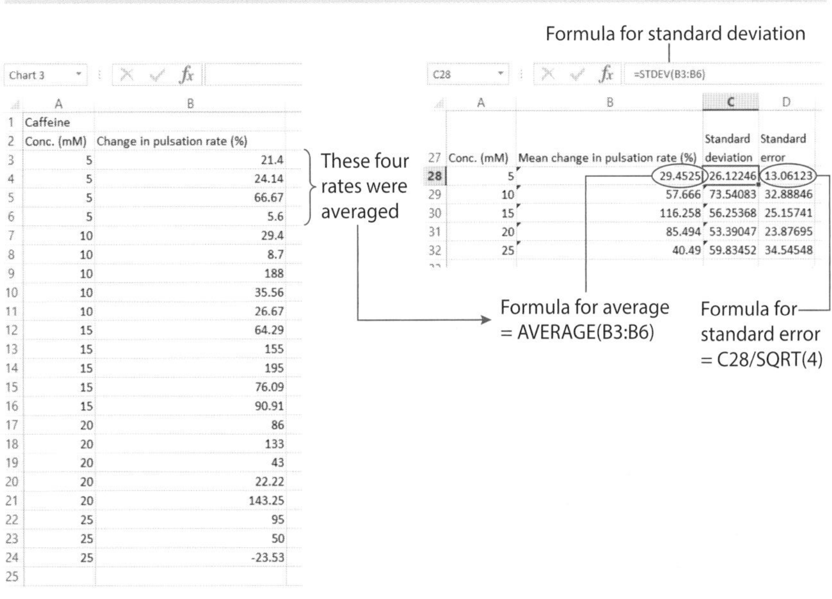

	A	B
1	Caffeine	
2	Conc. (mM)	Change in pulsation rate (%)
3	5	21.4
4	5	24.14
5	5	66.67
6	5	5.6
7	10	29.4
8	10	8.7
9	10	188
10	10	35.56
11	10	26.67
12	15	64.29
13	15	155
14	15	195
15	15	76.09
16	15	90.91
17	20	86
18	20	133
19	20	43
20	20	22.22
21	20	143.25
22	25	95
23	25	50
24	25	-23.53
25		

	A	B	C	D
27	Conc. (mM)	Mean change in pulsation rate (%)	Standard deviation	Standard error
28	5	29.4525	26.12246	13.06123
29	10	57.666	73.54083	32.88846
30	15	116.258	56.25368	25.15741
31	20	85.494	53.39047	23.87695
32	25	40.49	59.83452	34.54548

Figure A2.9 Calculation of mean, standard deviation, and standard error.

In this example, C28 is the cell that contains the value for standard deviation, SQRT is the formula for square root, and N is the number of measurement values that were averaged to calculate the mean.

3. Now click the chart area of the graph displaying the mean values. To open the Format Error Bars pane:

   Mac. Click **Chart Design | Add Chart Element | Error Bars | More Error Bar Options.**

   Windows. Click the + **Chart Elements** button and then **Error Bars | More Options.**

4. If you would like the vertical error bars to extend above and below the mean value, keep the default **Direction: Both**. Under **Error Amount**, click **Custom** and then **Specify Value** to open a second small **Custom Error Bars** dialog box. Click the red arrow for **Positive Error Value** and select the cells in your worksheet that contain the standard error values. Click the red arrow again to close the box. Repeat this process for **Negative Error Value.** Click **OK** to close the dialog box. In Windows, click **OK** to close the **Format Error Bars** dialog box.

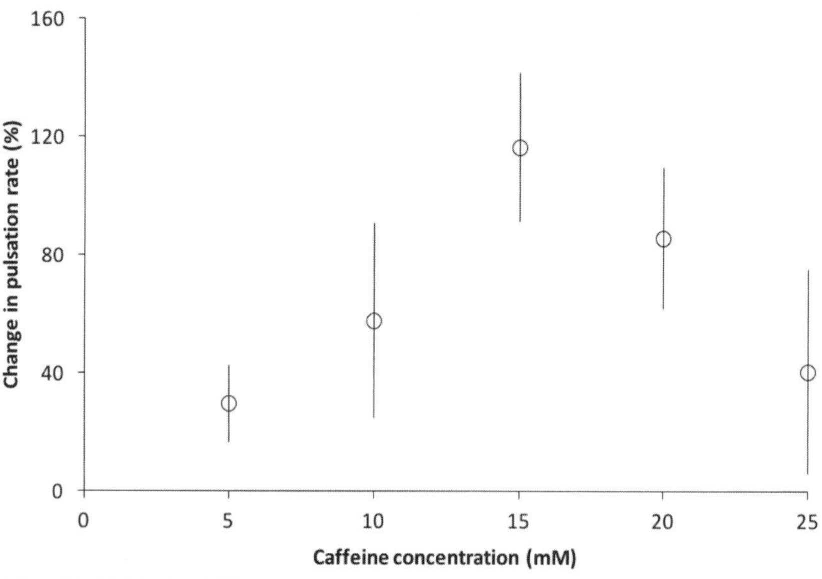

Figure A2.10 Vertical error bars inserted about the mean value to show variance.

5. If your version of Excel for some crazy reason has inserted horizontal error bars, delete them by clicking any one of them and pressing **Delete**. Your graph of the mean values with error bars will now look like Figure A2.10.

Data analysis with error bars

The addition of error bars to the mean values changes our interpretation of the results. The larger the standard error, the less confidence we have that the mean represents the true value. Furthermore, the more the error bars overlap, the less likely that these measurement values differ significantly from each other. Based on the mean values alone, we might have concluded that there was a real difference in pulsation rate with every 5 mM increase in caffeine concentration (see Figure A2.8B). The large overlap between the error bars for 5 and 10 mM caffeine, however, suggests that the pulsation rates do not differ that much for these concentrations (see Figure A2.10). On the other hand, the error bars for 10 and 15 mM caffeine barely overlap, so we are more confident that there may be a real difference in pulsation rate between these two concentrations.

Go to the **COMPANION WEBSITE** • **sites.sinauer.com/knisely5e**
for samples, template files, and tutorial videos

Appendix 3

Preparing Oral Presentations with Microsoft PowerPoint

Introduction

Microsoft PowerPoint allows you to create a slide show containing text, graphics, and audiovisual media. Besides being used as a visual aid during the presentation, the **slide deck** (the slides and any notes that make up a presentation) may be posted online as a study aid or printed out for taking notes. Slides may also be reused in other presentations and shared for documentation purposes, particularly in the corporate environment.

PowerPoint 2013 and 2016 for Windows and PowerPoint for Mac 2016 have Ribbon interfaces. The **Ribbon** is a single strip that displays **commands** in task-oriented **groups** on a series of **tabs** (Figure A3.1). In Windows, additional commands in some of the groups can be accessed with the **dialog box launcher**, a diagonal arrow in the right corner of the group label. In the Mac OS, additional commands are located on the **menu bar**. The **Quick Access Toolbar** comes with buttons for saving your file and undoing and redoing commands; you can also add buttons for tasks you perform frequently. The Ribbon interface has characterized the past several versions of Word, so finding commands in the latest version should not be difficult.

The first section of this appendix emphasizes the importance of planning your presentation before you even open PowerPoint. The planning stage is where you think about what your audience already knows and how to present new information in an understandable and engaging manner. Ideally, you want to give a presentation that is worth your audience members' time.

The second part of this appendix shows you how to construct slides that are audience-friendly. Audience-friendly slides are, first of all, easy on the eyes, and, secondly, focused on a single idea. "Easy on the eyes" means not just esthetically pleasing, but legible. Text must be large enough to be read from the back of the presentation room. Images must be sharp.

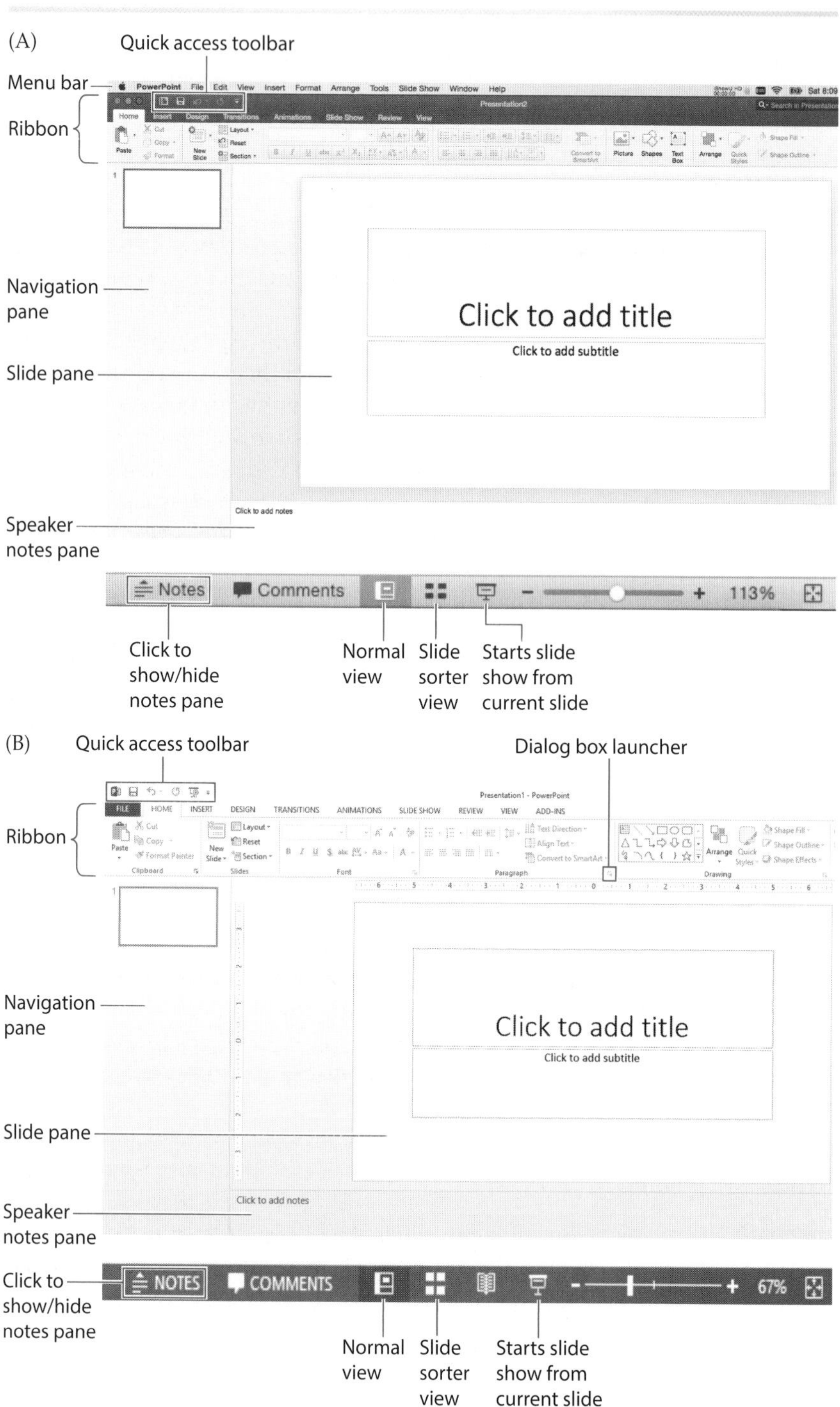
(A)
Quick access toolbar
Menu bar
Ribbon
Navigation pane
Slide pane
Speaker notes pane
Click to add title
Click to add subtitle
Click to add notes
Notes
Comments
113%
Click to show/hide notes pane
Normal view
Slide sorter view
Starts slide show from current slide
(B)
Quick access toolbar
Dialog box launcher
Ribbon
Navigation pane
Slide pane
Speaker notes pane
Click to add title
Click to add subtitle
Click to add notes
NOTES
COMMENTS
67%
Click to show/hide notes pane
Normal view
Slide sorter view
Starts slide show from current slide

◀ **Figure A3.1** Screen display of part of the command area in Normal view. (A) PowerPoint for Mac 2016 presentation, and (B) PowerPoint 2013 presentation. The status bar at the bottom of the screen has commands for changing views and showing and hiding the notes pane.

Both text and images must stand out on the background. Slides that focus audience attention on a single idea contain a minimum amount of text that, whenever possible, is supported with a visual aid. Animations can be used to emphasize one element of the slide at a time. Speaker prompts that potentially make slides text-heavy are entered in the Notes pane, which is seen by the speaker in Presenter View, but not by the audience.

The third part of this appendix is about delivering your PowerPoint presentation. For a delivery to be successful, the speaker must first practice reading and refining the script until he/she has practically memorized it. I will describe a new feature in PowerPoint called **Presenter View**, which displays the current slide seen by the audience, along with the next slide (or animation) and prompts in the Notes pane, which are seen only by the speaker. This presentation mode is useful for rehearsal as well as for keeping the speaker on track during the actual presentation. I will also describe the tools that a presenter needs to navigate among slides and to write on the slides. Finally, I will give examples of ways to share the slide deck beyond the presentation. For all other PowerPoint questions, please visit the Microsoft Office support website.

In this book, the nomenclature for the command sequence is as follows:

> **Ribbon tab | Group** (Windows only) | **Command button | Additional Commands** (if available).

For example, to insert a slide with title and content text boxes in Windows, **Home | Slides | New Slide ▼ | Title and Content** means "Click the **Home** tab and in the **Slides** group, click the down arrow (▼) on the **New Slide** button to select the **Title and Content** option. In the Mac OS, the corresponding command sequence is **Home | New Slide ▼ | Title and Content**.

Handling computer files

The section "Good Housekeeping" in Appendix 1 applies equally to Word documents and PowerPoint presentations. Read over this section to develop good habits for naming, organizing, and backing up computer files.

Planning Your Presentation

Before you open PowerPoint, make a concept map of your presentation (see p. 41 in Chapter 3). Think about what your audience already knows and how to lead them to what you want them to know. Very roughly sketch or outline the topics and decide on the order in which you want to present them. It doesn't matter if you use notebook paper, a whiteboard, or an electronic device—the important thing is to give your ideas some structure. The concept map doesn't have to be perfect. It is very common to change things around in the course of preparing a presentation.

As much as possible, make or find visuals that support the most important points of your talk. In biology talks, it is common to see photos, graphs, gel images, phylogenetic trees, and other visual aids. Make sure that these figures are large, legible, and sharp when projected onto a large screen. As you prepare the visuals, write down what you want to say about them.

After preparing the individual components of your presentation, you are finally ready to open PowerPoint, choose an appropriate slide design (theme), and insert content on the slides.

Designing Slides

Selecting a theme

When you open PowerPoint, you have a number of options. You can open an existing presentation, select a blank presentation or a colorful template or themed presentation from the gallery, or search for templates or themes online.

Mac **File | New Presentation** opens a blank presentation. **File | New from Template** opens the gallery of themes, which includes **Blank Presentation**.

Windows **File | New** opens the gallery of themes, which includes **Blank Presentation**.

Themes (designs) with a light background make the room much brighter during the actual presentation, which has the dual advantage of keeping the audience awake and allowing you to see your listeners' faces. Many professional speakers, however, prefer white text on a dark background because the text appears larger. Whatever your preference, choose a theme that reflects your style, is appropriate for the topic, and complements the content of the individual slides.

From the perspective of the audience, two non-negotiable criteria for slide design are **contrast** and **font size**. For content to be legible from a distance in a large room, the text must stand out from the background, and the font size has to be at least 24 pt. To determine if a theme meets these criteria:

Mac

1. Click inside one of the text boxes in the Slide pane (see Figure A3.1A). Click a theme on the **Design** tab. The down and right arrows display more themes. After clicking a theme, choose a color and background style from the options on the right. Keep in mind that black on a light background or white on a dark background provides the best contrast.
2. To check the default font sizes for this theme variant, click **View | Slide Master** and then scroll to the very first slide thumbnail in the navigation pane on the left. When you edit this slide, the changes will be applied to all of the slides in the presentation, not just to certain slide layouts. Click each style (title, first level text, second level text, and so on) to view the font style and size on the **Home** tab of the Ribbon. Triple-click to select any levels that are less than 24 pt and increase the font size accordingly.
3. To exit Slide Master mode, click the **Slide Master** tab and **Close Master**.

Windows

1. Click inside one of the text boxes in the Slide pane (see Figure A3.1B). Click the **Design | Themes | More** arrow to display all of the themes. Mouse over the themes. The Live Preview feature allows you to see how the change affects the slide before actually applying it. After you click a theme, mouse over the color palettes and background styles in the **Variants** group. Keep in mind that black on a light background or white on a dark background provides the best contrast.
2. To check the default font sizes for this theme variant, click **View | Master Views | Slide Master** and then scroll to the very first slide thumbnail in the navigation pane on the left. When you edit this slide, the changes will be applied to all of the slides in the presentation, not just to certain slide layouts. Click each style (title, first level text, second level text, and so on) to view the font style and size under **Home | Font**. Triple-click to select any levels that are less than 24 pt and increase the font size accordingly.
3. To exit Slide Master mode, click the **Slide Master** tab and **Close Master**.

Choosing a slide size

Almost all TVs, computer screens, and projection screens have an aspect ratio of 16:9 (widescreen), which is why widescreen is the default slide size in PowerPoint 2013 and 2016 and PowerPoint for Mac 2016. Widescreen, as the name implies, increases the width of the slide, so you have more room for content. However, if a slide formatted in widescreen is projected in the older 4:3 format, the sides of the slide will be cut off. For this reason, it's a good idea to find out beforehand how the slides at the presentation site will be displayed. If necessary, the slide size can be changed in PowerPoint by clicking **Design | Customize** (Windows only) | **Slide Size ▼**.

Views

The most frequently used views can be accessed by clicking a button on the status bar at the bottom of the screen (see Figure A3.1). Most of the time, you will be working in **Normal** view. **Normal** view consists of a Navigation pane, a Slide pane whose layout and content can be customized, and a Notes pane where you can write notes for your presentation, which will not be seen by the audience. The Notes pane can also be used to provide explanatory information about each slide if the slide deck will be shared. There is no limit to the amount of text that can be entered on the Notes pane.

Slide Sorter view is useful for evaluating the overall appearance of a presentation because it displays thumbnails of all of the slides in the presentation, complete with slide transitions. You cannot edit the content of an individual slide in **Slide Sorter** view, but you can rearrange the slides. To select a slide to move, copy, or delete, single-click it. To edit the content of the slide, double-click it to return to **Normal** view. Slides can also be rearranged in the Navigation pane in **Normal** view.

Slide layouts

Presentations in biology usually follow the Introduction-Body-Closing format. In the Introduction, the speaker guides the audience from information that is generally familiar to them to new information, which is the focus (Body) of the talk. The take-home message for the audience and any acknowledgments are given in the Closing.

Title slide The default first slide in a new presentation is the Title Slide (see Figure A3.1). Follow the instructions in the placeholders to add text or add custom content. The same content will appear on the slide thumbnails on the left.

The Title Slide is going to set the tone for your presentation. A catchy title and interesting visuals are likely to capture your audience's attention. If possible, allude to the main benefit your audience will gain from listening to your presentation.

Adding slides To insert a slide after the title slide, click below the slide thumbnail in the Navigation pane. A red horizontal line shows where the next slide will be inserted. Click **Home | Slides** (Windows only) | **New Slide ▼** and choose a slide layout. The name of the dialog box is the theme you selected for your presentation. The theme for a blank presentation is Office Theme.

The basic slide layout options allow you to arrange text and content (pictures, charts, tables, videos, and graphics) in one or two columns. The placeholder text boxes, however, can be moved, resized, and deleted as needed. Alternatively, if you decide that a different layout would work better after you've already added content to a slide, click **Home | Slides** (Windows only) | **Layout ▼** and select a different layout.

The last slide A professional slide show has a definitive ending. End the show with an acknowledgments slide or add a slide that invites the audience to ask questions.

Reusing slides from other presentations

Slides from other presentations can be inserted in a new presentation. In Windows, the theme from the reused slide can be applied to other slides in the new presentation, or the reused slide can adopt the theme of the new presentation.

Mac

1. In the Navigation pane of the presentation you are working on, click where you want the reused slides to be inserted. This location will be marked by a horizontal red line.
2. To import *all of the slides* in the source presentation, click **Insert | New Slide ▼ | Reuse Slides** and navigate to a PowerPoint presentation you want to reuse. Click **OK**. The imported slides will automatically adopt the theme of the target presentation.
3. To import *only certain slides* from another presentation, open that presentation, and click the slides to copy while holding down the ⌘ key. Click **Home | Copy** or press **⌘+C**. In the target presentation, click **Home | Paste** or press **⌘+V** to apply the target theme to the reused slides. Click **Home | Paste ▼ | Keep Source Formatting**

if you intend to change the theme of the target presentation to that of the reused slides.

Windows

1. In the presentation you are working on, click **Insert | New Slide ▼ | Reuse Slides**. In the Reuse Slides pane on the right, navigate to a PowerPoint presentation you want to open. Thumbnails of the slides in the selected presentation will then be shown in the Reuse Slides pane.
2. In the Navigation pane on the left, click where you want the reused slide to be inserted. This location will be marked by a horizontal red line. In the Reuse Slides pane on the right, click the thumbnail of the slide to be reused.
3. The theme of the presentation you are working on will be applied to this slide unless you click the **Keep source formatting** checkbox at the bottom of the Reuse Slides pane. If you prefer to apply the theme of the reused slide to your current presentation, right-click the reused slide and select **Apply Theme to All Slides**.

Adding animation

Without animation, all of the objects on a slide are displayed for as long as the slide is displayed. **Animation** causes individual objects on a slide to move. Speakers use animation to help the audience focus on one element at a time, which is only revealed when the speaker is ready to talk about it. This technique is often used to list objectives or summarize conclusions.

If you decide to use animations, use the same animation throughout your presentation and use it selectively for emphasis. *You want your audience to be focused on what you are saying, not be distracted by the special effects!*

Mac To animate an object, click it, and then select an animation from the **Animations** tab. Click the respective down arrows for the entrance and exit animations to display categories with different levels of excitement. For a scientific audience, it's probably best to stick to the basic animations, not just because they appear and disappear quickly.

To adjust the direction and sequence of the animation effect, click the **Effect Options** button. Other commands on the **Animations** tab allow you to customize how the animation will start (on click or connected to the previous animation), the duration, and delay time. To preview your animation, click the **Slide Show** button on the right side of the status bar (see Figure A3.1A). To delete an animation, click **Animations | Animation Pane**, select the animation to delete, and click the red X. To change the order in which the animations are played, use the up and down arrows on the Animation pane.

Windows To animate an object, click it, and then select an animation from the **Animations** tab. Click the **More** arrow next to the animations and preview the options by clicking the respective buttons. Notice that there are both entrance and exit animations. More effects, with increasing levels of excitement, can be selected at the bottom of the options box. For a scientific audience, it's probably best to stick to the basic animations, not just because they appear and disappear quickly.

To adjust the direction and sequence of the animation effect, click the **Effect Options** button. Other commands on the **Animations** tab allow you to customize how the animations will start (on click or connected to the previous animation), the duration, delay time, and order in which they will be played. To delete an animation, click it and choose **Animations | Animation | None**. To preview your animation, click the **Slide Show** button on the right side of the status bar (see Figure A3.1B).

Importing Excel graphs

Excel charts can be imported into PowerPoint with full editing capability using the standard copy and paste commands. However, the drawback is that Excel's default 10 pt font makes the numbers and text close to unreadable for the audience. If you don't need to edit the chart in PowerPoint, then a better option is to paste the graph as a picture. To do so in the Mac OS, click **Home | Paste ▼ | Paste as Picture**. To do so in Windows, click **Home | Paste ▼ | Picture**. The font size automatically increases when you drag on one of the corners to enlarge the picture.

Working with shapes

Microsoft PowerPoint and Word offer a wide variety of shapes with which to make pointers, line drawings, and simple graphics. You can also make unique shapes by combining, coloring, and editing preexisting shapes. Begin by clicking **Insert | Illustrations** (Windows only) **| Shapes** and selecting a shape.

Lines Lines (including arrows) are useful as pointers. After clicking a line from the shapes gallery and moving the mouse over the Slide pane to display crosshairs, hold down the left mouse button, drag to where you want the line to end, and release the mouse button. To make perfectly horizontal or vertical lines, hold down the **Shift** key while holding down the left mouse button and drag. In general, when it's important that a shape retain perfect proportions (e.g., circle not oval; square not rectangle), hold down the **Shift** key while drawing or resizing the shape with the mouse.

To change the angle or length of the line, click the line and then position the mouse pointer over one of the end points of the line to display a two-headed arrow. Hold down the left mouse button, drag to where you want the line to end, and release the mouse button.

To format the line, click it and select an attribute on the **Shape Format** tab in the Mac OS or **Drawing Tools Format** tab in Windows. Customize line color, weight, dash type, and arrow type, and, if you want, add special effects to give the line a three-dimensional look.

Text To add text to your drawing, click **Insert | Text** (Windows only) **| Text Box**. Position the mouse pointer where you want the text box to appear, hold down the left mouse button, drag to enlarge the text box to the size you think you'll need (you can resize it later), and release the mouse button. Type text inside the box. In the Mac OS, change the font face and size by selecting the text and then changing the attributes on the **Home** tab. In Windows, right-click the text and use the buttons on the mini toolbar. Or add special effects using the buttons on the **Shape Format** tab in the Mac OS or **Drawing Tools Format** tab in Windows.

Text can also be added to two-dimensional shapes, such as circles, triangles, and squares. With the shape selected, just start typing. To format the text, use the options on the **Home** tab or on the **Shape Format** tab (Mac) or the **Drawing Tools Format** tab (Windows).

Group shapes When several objects make up a graphic, it may make sense to group them as a unit. Grouping allows you to copy, move, format, and animate all the objects in the graphic at one time. To group individual objects into one unit, click each object while holding down the **Shift** key.

- **Mac** Release the **Shift** key and then click **Group Objects** on the **Shape Format** tab. One set of selection handles surrounds the entire unit when the individual objects are grouped. To ungroup, simply select the group and click **Group | Ungroup**. Grouping is temporary, whereas merging is permanent (see the next section).
- **Windows** Release the **Shift** key and then click **Group** under **Drawing Tools Format | Arrange**. One set of selection handles surrounds the entire unit when the individual objects are grouped. To ungroup, simply select the group and click **Group | Ungroup**. Grouping is temporary, whereas merging is permanent (see the next section).

Merge shapes This command makes it possible to make unique shapes by permanently grouping standard shapes. For example, if you created a cartoon of a newly discovered organism, and you want to reuse the entire cartoon, not just the individual shapes, shift-click the individual objects. In the Mac OS, click **Shape Format | Merge Shapes**. In Windows, click **Drawing Tools Format | Insert Shapes | Merge Shapes**. *Note that once a presentation containing a merged shape has been saved, the shape cannot be unmerged.*

Align shapes When there are multiple objects on a slide, alignment guides appear when you drag one of the objects. **Alignment guides** are vertical and horizontal lines that show you the edges of other objects with which you may wish to align the selected object.

Alternatively, you can use the **Align** button to align selected objects by their top, bottom, right, or left edges or center them vertically or horizontally. To do so, shift-click each object that you want to line up. Release the **Shift** key, and then:

- **Mac** Click **Shape Format | Align ▼**, and select one of the alignment options.
- **Windows** Click **Drawing Tools Format | Arrange | Align ▼**, and select one of the alignment options.

Finally, to move objects just a fraction of a millimeter from their current position, click the object, hold down the ⌘ key (Mac) or the **Ctrl** key (PC), and use the arrow keys to nudge the object exactly where you want it on the slide.

Adding videos

Inserting and playing videos in the new versions of PowerPoint are much easier than they were in previous versions. To insert a video file on your slide:

- **Mac** Click **Insert | Video** inside a text box. Choose either **Movie Browser** or **Movie from File** to insert a video that is saved on your computer. An add-in is required to link online videos to your presentation (see "Add a video to your PowerPoint 2016 for Mac presentation" on the Microsoft Office support website).
- **Windows** Click **Insert | Media | Video**, and choose either **Online Video** or **Video on My PC**. When using online videos, you have the choice of inserting the actual video file or embedding the code on your slide. Embedding reduces the size of your PowerPoint presentation, but requires a reliable Internet connection to play the video. To embed a YouTube video:
  1. Go to www.youtube.com and find the desired video.
  2. Below the video frame, click **Share** and then **Embed**. Copy the embed code.
  3. Paste the embed code into the text box in PowerPoint. Click ⇨ to exit the dialog box. A black box will be inserted on the slide, which you can size and reposition as needed.
  4. To preview the video, right-click it and select **Preview** from the drop down menu.

For more information on inserting and playing videos in PowerPoint 2013, see http://www.gcflearnfree.org/powerpoint2013/inserting-videos/1/ or the Microsoft Office support website.

Text formatting shortcuts

Mac To insert symbols (Greek letters, mathematical symbols, pictographs, etc.), click **Insert | Symbol**. Click a category (or type "Greek" into the search box) and double-click the symbol you want. Exit the dialog box. The symbol will be set in the same font as the text font in your PowerPoint presentation.

In PowerPoint, it is not possible to make shortcut keys for symbols, but it is possible to program them in AutoCorrect. To do so, copy the symbol and then click **Tools | AutoCorrect**. On the **AutoCorrect** tab, paste the symbol into the **With** box (it is not automatically entered, as it is in Word). Type a simple keystroke combination in the **Replace** box. Click the + sign at the bottom of the box, and then exit AutoCorrect.

A long expression such as 2-nitrophenyl β-D-galactopyranoside can also be replaced with a shorter one using AutoCorrect, but it is still not possible to italicize scientific names of organisms automatically as it is in MS Word's AutoCorrect.

Windows To insert symbols (Greek letters, mathematical symbols, Wingdings, etc.), click **Insert | Symbols | Symbol**, and click the desired symbol and close the box. For a uniform look, select the symbol on your PowerPoint slide and apply the text font that is used in your PowerPoint presentation.

In PowerPoint, it is not possible to make shortcut keys for symbols, but it is possible to program them in AutoCorrect. To do so, copy the symbol and then click **File | Options | Proofing | AutoCorrect Options**. On the **AutoCorrect** tab, paste the symbol into the **With** box (it is not automatically entered, as it is in Word). Type a simple keystroke combination in the **Replace** box and then exit AutoCorrect.

Interestingly, many symbols have already been entered on the **MathAutoCorrect** tab, but these keystrokes only work inside equation boxes. However, you can copy the symbols and their corresponding keyboard shortcuts from the **MathAutoCorrect** tab and paste them into their respective boxes on the **AutoCorrect** tab.

A long expression such as 2-nitrophenyl β-D-galactopyranoside can also be replaced with a shorter one using AutoCorrect, but it is still not possible to italicize scientific names of organisms automatically as it is in MS Word's AutoCorrect.

TABLE A3.1 Presentation tips

Do	Don't
Keep the wording simple	Write every word you're going to say on the slide
Make the text large and legible	Use backgrounds that make text and images hard to read
Strive for a consistent look (use the same font and format for each slide)	Use distracting animations and slide transitions, or sound effects
Include visuals that complement and support the auditory information	Include tables when you can show the trend better with a graph
Allow enough time for each slide	Talk endlessly without referring to a visual
	Rush through the slides without mentioning the take-home message about each one

Delivering Presentations

A PowerPoint presentation can be "delivered" electronically by attaching it to an email or posting it to a website, but most presentations are delivered in person. Table A3.1 gives some presentation tips. See Chapter 8 for detailed instructions on planning and delivering oral presentations.

Presenter View

Presenter View is a new feature in PowerPoint 2013 and 2016 as well as PowerPoint for Mac 2016 (Figure A3.2). This view is seen only by the presenter. It helps the speaker stay on track by displaying speaking prompts and a preview of the next slide without making the slide that the audience sees text-heavy. Audience members see only the current slide with the content that you specifically chose to help them focus on the main point.

Rehearsal

Mac To use Presenter View on your computer when you are rehearsing your presentation, press **Option+Return** on the keyboard or click **Slide Show | Presenter View** on the Ribbon. As shown in Figure A3.2A, there is a timer on the left and a clock on the right above the current slide. The timer starts as soon as you start Presenter View, but it can be paused or reset as needed. Notes can be added in this view, but if you need to edit a slide, press **Esc** to exit. Changes to slides can only be made in **Normal** view.

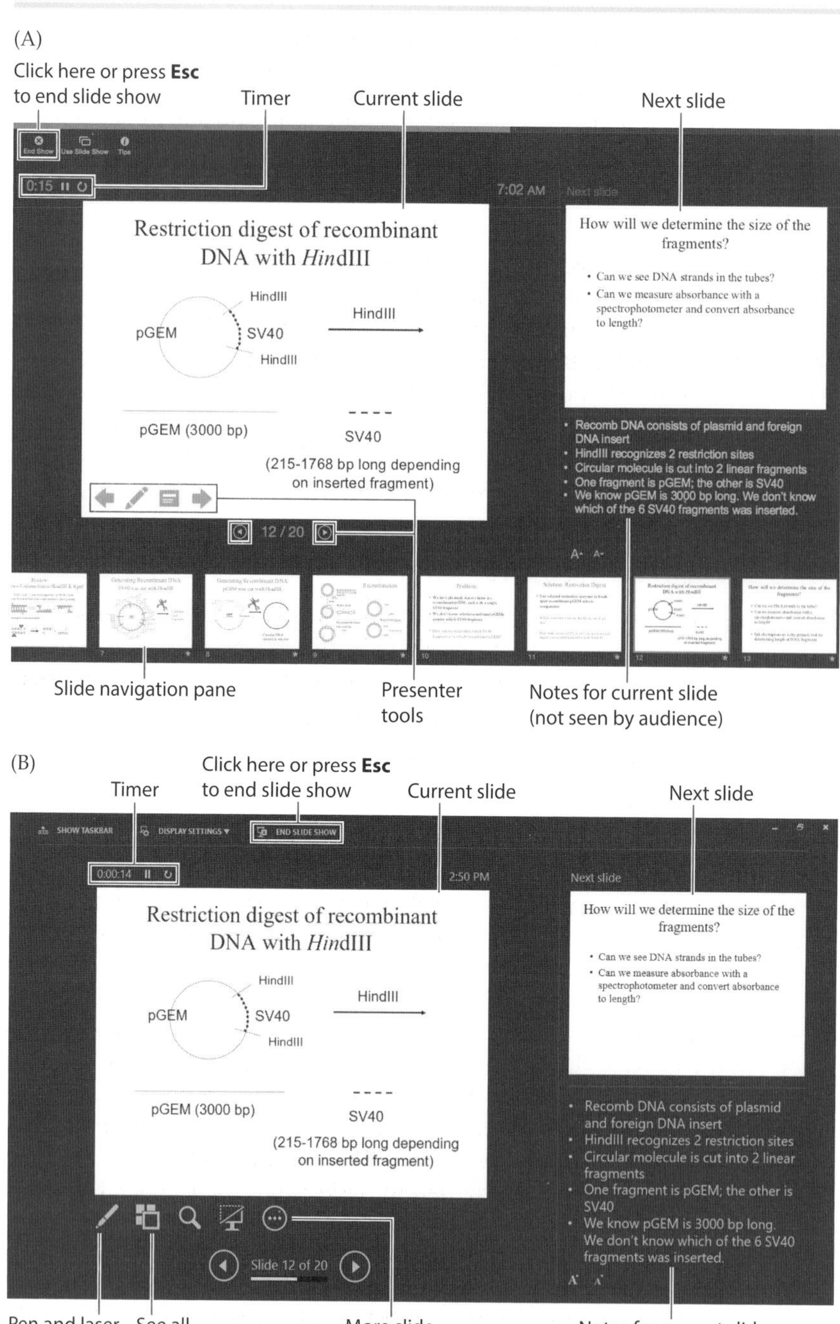
(A)
Click here or press **Esc** to end slide show
Timer
Current slide
Next slide
End Show
Use Slide Show
Tips
0:15
7:02 AM
Next slide
Restriction digest of recombinant DNA with *Hin*dIII
HindIII
pGEM
SV40
HindIII
HindIII
pGEM (3000 bp)
SV40
(215-1768 bp long depending on inserted fragment)
12 / 20
How will we determine the size of the fragments?
Can we see DNA strands in the tubes?
Can we measure absorbance with a spectrophotometer and convert absorbance to length?
Recomb DNA consists of plasmid and foreign DNA insert
HindIII recognizes 2 restriction sites
Circular molecule is cut into 2 linear fragments
One fragment is pGEM; the other is SV40
We know pGEM is 3000 bp long. We don't know which of the 6 SV40 fragments was inserted.
Slide navigation pane
Presenter tools
Notes for current slide (not seen by audience)
(B)
Timer
Click here or press **Esc** to end slide show
Current slide
Next slide
SHOW TASKBAR
DISPLAY SETTINGS
END SLIDE SHOW
0:00:14
2:50 PM
Next slide
Restriction digest of recombinant DNA with *Hin*dIII
HindIII
pGEM
SV40
HindIII
HindIII
pGEM (3000 bp)
SV40
(215-1768 bp long depending on inserted fragment)
How will we determine the size of the fragments?
Can we see DNA strands in the tubes?
Can we measure absorbance with a spectrophotometer and convert absorbance to length?
Recomb DNA consists of plasmid and foreign DNA insert
HindIII recognizes 2 restriction sites
Circular molecule is cut into 2 linear fragments
One fragment is pGEM; the other is SV40
We know pGEM is 3000 bp long. We don't know which of the 6 SV40 fragments was inserted.
Slide 12 of 20
Pen and laser pointer tools
See all tools
More slide show options
Notes for current slide (not seen by audience)

◀ **Figure A3.2** Presenter View is seen only on the presenter's computer screen. This view shows, on the left, the current slide that is seen by the audience; speaker notes for that slide bottom right; and the next slide top right. Presenter tools allow the speaker to write on the slides, turn the mouse into a laser pointer (Windows only), navigate among slides, zoom in, and black out the screen. (A) PowerPoint for Mac 2016 screen shot. (B) PowerPoint 2013 screen shot.

Windows To use Presenter View on your computer when you are rehearsing your presentation, click **Alt+F5** on the keyboard or click **Slide Show | Start Slide Show** on the Ribbon and then right-click and select **Show Presenter View** from the menu. As shown in Figure A3.2B, there is a timer on the left and a clock on the right above the current slide. The timer starts as soon as you start Presenter View, but it can be paused or reset as needed. If you need to edit a slide or the notes, press **Esc** to exit. Changes can only be made in **Normal** view.

Resources at the presentation site

To deliver a PowerPoint slide show, you'll need a computer and projection equipment as well as a screen on which to display the slides. Assuming these requirements are met, there are several ways to run the presentation:

- From the hard drive of your laptop computer
- From a flash drive on a computer at the presentation site
- From a Web server

If you are using your own laptop, connect it to the projector, start PowerPoint, and open the presentation. Run through the presentation to make sure the animations and videos work. Close unnecessary programs that might cause the computer to run slower or the audience to become distracted.

Mac To start a presentation in **Slide Show mode**, click the **Slide Show** button in the lower right corner of the screen (see Figure A3.1A). If you exit Slide Show mode and then click the **Slide Show** button again, the slide show resumes at the current slide. To start the show at the beginning, press **⌘+Shift+Return** or click **Slide Show | Play from Start**.

To switch to Presenter View when in Slide Show mode, click the **Slide Show options** button on the presenter toolbar (Figure A3.3), and select **Use Presenter View** from the menu.

Windows To start a presentation in **Slide Show mode**, click the **Slide Show** button in the lower right corner of the screen (see Figure A3.1B). If you exit Slide Show mode and then click the **Slide Show** button again,

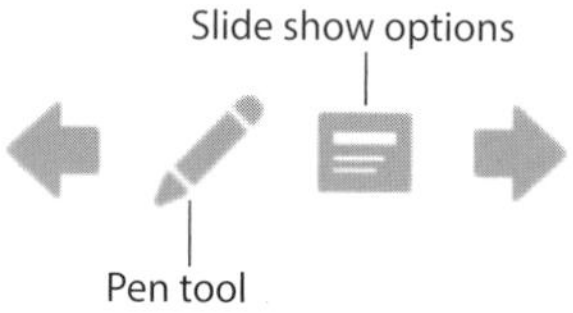

Figure A3.3 Presenter toolbar in the Mac OS appears at the bottom of the current slide in Slide Show mode and Presenter View. The Slide Show options button opens a menu that lets you navigate to another slide, make the screen black or white to focus audience attention away from the projection screen, end the slide show, and toggle between Slide Show mode and Presenter View. The Pen tool allows you to write on the slides. The writing is temporary and will not be saved after you navigate to another slide.

the slide show resumes at the current slide. **Shift+F5** is the corresponding keyboard shortcut. To start the show at the beginning, press **F5** or click **Slide Show | Start Slide Show | From Beginning**.

To switch to Presenter View when in Slide Show mode, right-click and select **Show Presenter View** from the menu.

If the presentation site has a computer, but it is not hooked up to the Internet, save your presentation file and any linked videos on a flash drive and carry it with you. Assuming that PowerPoint is installed on the presentation room's computer, plug the flash drive into the USB port, download the files to desktop, and start your presentation as described previously.

To run a presentation from a Web server, first find out if you have access to the Internet at the presentation site. If so, save your file to a remote server, such as Google Drive or Dropbox. Similarly, if your college or university has networked computers, you can prepare your presentation on a networked computer in your room, save the file in your Netspace, and then access the file from another networked computer in the room where you will hold your talk.

Presenter tools

When you give your presentation, some of the actions you will perform are

- Starting and ending your slide show
- Moving forward through the slides
- Navigating to selected slides
- Pointing to specific slide elements
- Writing on the slides for emphasis

PowerPoint offers many presenter tools that can be accessed by keyboard and/or mouse on the computer that is connected to the projector. To use these tools, however, you have to be standing near the computer.

If you prefer to move around during your talk, a wireless presenter allows you to change slides from anywhere in the room. This device usually consists of a USB receiver that you plug into the computer's USB port, and the remote control, which has buttons for moving forward and backward, pausing, or stopping the slide show. The remote control may also have a built-in laser pointer.

Mac To advance to the next slide:

- On the keyboard, press **Return** or the space bar.
- Using the mouse, click the forward arrow in the toolbar or click the left mouse button. In Presenter View, you can also click the current slide.

If a slide has animations, these procedures will advance to the next animation. To go back to the previous slide or animation:

- Press **Delete**.
- Click the back arrow in the toolbar.

To navigate to a specific slide in Slide Show mode, click **Slide Show options | By Title**, and select the slide from the menu. In Presenter View, click a slide from the Slide Navigation pane at the bottom of the screen (see Figure A3.2A). In either mode, if you happen to know the number of the slide, type the number and hit **Return**.

To write on the slides, click the **Pen tool**.

To end a slide show, press **Esc** or click the **End Show** button at the top of the Presenter View window.

Windows To advance to the next slide:

- On the keyboard, press **Enter** or the space bar.
- Using the mouse, click the forward arrow in the toolbar, click the left mouse button, or click either the current or the next slide.

If a slide has animations, these procedures will advance to the next animation. To go back to the previous slide or animation:

- Press **Backspace**.
- Click the back arrow in the toolbar.
- Click the right mouse button to display a menu of navigation options.

To navigate to a specific slide in Slide Show mode, right-click and select **See all slides** from the menu. In Presenter View, click the **See all slides** button in the presenter toolbar at the bottom of the screen (see Figure A3.2B) and then click the desired slide. In either mode, if you happen to know the number of the slide, type the number and hit **Enter**.

To turn the mouse into a laser pointer or to write on the slides, click **Pen and laser pointer tools**.

To end a slide show, press **Esc** or click the **End Slide Show** button at the top of the Presenter View window.

Speaker notes, handouts, and sharing electronic presentations

The Notes pane is useful for listing the talking points for each slide and for providing explanatory information in handouts and for other presenters who may be reusing your slide show. If you are comfortable referring to your notes on a computer screen in Presenter View, then there is no need to print out speaker notes. However, if you feel more comfortable presenting with notes in hand, follow the instructions below to print a large thumbnail of the slide along with notes. When printing on a black-and-white printer, select **Pure Black and White** to avoid printing out the background of the slides. The same instructions can be used to produce printed handouts or PDFs if posting online.

Mac Click **File | Print | Show Details | Layout: Notes**.

Windows Click **File | Print | Print Layout | Notes Pages** or **File | Export | Create Handouts**.

Because high resolution images can increase the size of a PowerPoint presentation significantly, you may wish to compress the pictures first as follows:

Mac Click **Picture Format | Compress Pictures** and then choose an option from the menu.

Windows Click **Picture Tools Format | Adjust | Compress Pictures** and then choose an option from the menu.
Other ways to share slide decks are described in Table 8.2.

Bibliography

Databases and scholarly search engines

Harzing A-W, Alakangas S. 2016. Google Scholar, Scopus and the Web of Science: a longitudinal and cross-disciplinary comparison. Scientometrics 106: 787–804 doi:10.1007/s11192-015-1798-9

Hodge DR, Lacasse JR. 2013. Ranking Disciplinary Journals with the Google Scholar H-index: A New Tool for Constructing Cases for Tenure, Promotion, and Other Professional Decisions. Florida State University Libraries. [accessed 2016 Oct 13]. https://fsu.digital.flvc.org/islandora/object/fsu:252709/datastream/PDF/view

Michigan State University Libraries Research Guides. c2016. PubMed, Web of Science, or Google Scholar? A behind-the-scenes guide for life scientists [accessed 2016 Oct 13]. http://libguides.lib.msu.edu/pubmedvsgooglescholar

Moed HF, Bar-Ilan J, Halevi G. 2016. A new methodology for comparing Google Scholar and Scopus. *Journal of Informetrics* 10: 533–551

Evaluating web pages

Alexander J, Tate M. c1996-2005. Evaluate Web Pages. Chester (PA): Widener University; [accessed 2016 Oct 14]. http://www.widener.edu/about/campus_resources/wolfgram_library/evaluate/

Evaluating Resources. c2014-2016. Berkeley (CA): Regents of the University of California. [updated 2016 Aug 16; accessed 2016 Oct 14]. http://www.lib.berkeley.edu/TeachingLib/Guides/Internet/Evaluate.html

Olin & Uris Libraries: Critically Analyzing Information Sources. c2016. Ithaca (NY): Cornell University Library. [updated 2016 May 27; accessed 2016 Oct 14]. http://olinuris.library.cornell.edu/ref/research/skill26.htm

Purdue Online Writing Lab (OWL). c1995-2016. West Lafayette (IN): The Writing Lab & The OWL at Purdue and Purdue University. [updated 2013 Feb 22; accessed 2016 Oct 14]. http://owl.english.purdue.edu/owl/resource/553/01/

Internet sources, citing

Hume-Pratuch J. 2014. How to Use the New DOI Format in APA Style. APA Style Blog. [accessed 2016 Dec 31]. http://blog.apastyle.org/apastyle/2014/07/how-to-use-the-new-doi-format-in-apa-style.html

Patrias K. 2007. Citing medicine: the NLM style guide for authors, editors, and publishers. 2nd ed. Wendling DL, technical editor. Bethesda (MD): National Library of Medicine (US); [updated 2015 Aug 11; accessed 2016 Dec 31]. http://www.nlm.nih.gov/citingmedicine

Learning support

Bucknell University. c2016. The Teaching & Learning Center, Student Learning Support http://www.bucknell.edu/LearningCenter

Novak JD, Cañas AJ. 2008. The Theory Underlying Concept Maps and How to Construct and Use Them, Technical Report IHMC CmapTools 2006-01 Rev 01-2008, Florida Institute for Human and Machine Cognition; [updated 2008 Jan 22; accessed 2016 Nov 2]. http://cmap.ihmc.us/docs/pdf/TheoryUnderlyingConceptMaps.pdf

Palmer-Stone D. 2001. Learning Module: Reading and Concept Mapping. Victoria (BC), Canada: University of Victoria, Counselling Services; [accessed 2016 Dec 31]. http://www.uvic.ca/services/counselling/resources/learninghandouts/index.php

University of Victoria. c2016. Counselling Services Learning Handouts. Victoria (BC), Canada; [updated Mar 2010; accessed 2016 Nov 1]. https://www.uvic.ca/services/counselling/resources/learninghandouts/index.php

Microsoft Office

Apple Inc. c2016. Backing up your Mac hard drive. [accessed 2016 Dec 18]. https://www.apple.com/support/backup/

Hoffman C. c2016. What's the best way to back up my computer? [2016 Feb 29; accessed 2016 Dec 18]. http://www.howtogeek.com/242428/whats-the-best-way-to-back-up-my-computer/

Microsoft Support. c2016. Office Help and Training. [accessed 2016 Dec 18]. https://support.office.com/en-us

Miscellaneous

Light RJ. 2001. Making the most of college: Students speak their minds. Cambridge, MA: Harvard University Press. 242 p.

Martin B. 2011. Doing good things better. Ed (Sweden): Irene Publishing. [accessed 2016 Nov 16]. Available free online at: http://www.bmartin.cc/pubs/11gt/

Shaffer S. 2016. Active reading. iStudy for Success. ITS Training Services. [updated 2016 Jan 13; accessed 2016 Oct 25]. http://tutorials.istudy.psu.edu/activereading/

Oral presentations

Alley M. c2016. Assertion-Evidence Approach. State College (PA): The Pennsylvania State University; [accessed 2016 Oct 2]. http://www.craftofscientificpresentations.com/assertion-evidence-approach.html

D'Arcy J. 1998. Technically Speaking: A Guide for Communicating Complex Information. Columbus: Battelle Press. 270 p.

Fegert F, Hergenröder F, Mechelke G, Rosum K. 2002. Projektarbeit: Theorie und Praxis. Stuttgart: Landesinstitut für Erziehung und Unterricht Stuttgart. 226 p.

Hailman JP, Strier KB. 2006. Planning, proposing, and presenting science effectively: A Guide for Graduate Students and Researchers in the Behavioral Sciences and Biology, 2nd ed. Cambridge: Cambridge University Press. 248 p.

McConnell S. 2011. Designing effective scientific presentations. [uploaded 13 Jan 2011; accessed 2016 Dec 18]. https://www.youtube.com/watch?v=Hp7Id3Yb9XQ

Nathans-Kelly T, Nicometo CG. 2014. Slide Rules: Design, Build, and Archive Presentations in the Engineering and Technical Fields [e-book]. Hoboken (NJ): Wiley-IEEE Press. 219 p.

Plagiarism

Frick T. c2002–2016. How to recognize plagiarism: tutorials and tests. Bloomington (IN): Indiana University, School of Education; [updated 2016 June 7; accessed 2016 Oct 1]. https://www.indiana.edu/~academy/firstPrinciples/

Posters

Alley M. c1997-2016. Scientific Posters. State College (PA): The Pennsylvania State University; [accessed 2016 Sept 25]. http://www.craftofscientificposters.com/

Cothran K. c2009. Research Poster Design, Construction, and Printing. [accessed 2016 Sept 25]. http://www.clemson.edu/centers-institutes/crca/documents/templates/poster-design-presentation.pdf

Everson K. The scientist's guide to poster design. [accessed 2016 Sept 25]. http://www.kmeverson.org/academic-poster-design.html

Hess GR, Tosney K, Liegel L. c2014. Creating Effective Poster Presentations. [accessed 2016 Sept 25]. http://www.ncsu.edu/project/posters

Poster Creation and Presentation. c2016. State College (PA): The Pennsylvania State University; [last updated 2016 June 27; accessed 2016 Sept 25]. http://guides.libraries.psu.edu/c.php?g=435651&p=2973014

Purrington CB. c1997–2016. Designing conference posters. [accessed 2017 Jan 3]. http://colinpurrington.com/tips/poster-design

Woolsey JD. 1989. Combating poster fatigue: How to use visual grammar and analysis to effect better visual communications. *Trends in Neuroscience* 12(9): 325–332.

Reading and writing about biology

Gillen CM. 2007. Reading Primary Literature. San Francisco: Pearson/Benjamin Cummings. 44 p.

Hofmann AH. 2014. Scientific Writing and Communication: Papers, Proposals, and Presentations, 2nd ed. New York: Oxford University Press. 728 p.

Lannon JM, Gurak LJ. 2011. Technical Communication, 12th ed. White Plains (NY): Longman. 784 p.

McMillan VE. 2012. Writing Papers in the Biological Sciences, 5th ed. Boston: Bedford/St. Martin's. 256 p.

Pechenik JA. 2016. A Short Guide to Writing about Biology, 9th ed. Boston: Pearson Education. 262 p.

VanAlstyne JS. 2005. Professional and technical writing strategies, 6th ed. New York, NY: Pearson Longman. 752 p.

Reading journal articles

University of Minnesota Libraries. 2014. How to read and comprehend scientific research articles. [accessed 8 Nov 2016]. https://www.youtube.com/watch?v=t2K6mJkSWoA

Lockman T. 2012. How to read a scholarly journal article. [accessed 8 Nov 2016] https://www.youtube.com/watch?v=EEVftUdfKtQ

Revision

Corbett PB. 2011. The Reader's Lament. New York: The New York Times [accessed 2016 Nov 5]. http://afterdeadline.blogs.nytimes.com/2011/10/04/the-readers-lament/

Every B. c2006–2015. My 15 Best Proofreading Tips. BioMedical Editor [accessed 2016 Nov 5]. http://www.biomedicaleditor.com/proofreading-tips.html

LR Communication Systems, Inc. c1999. Proofreading and editing tips: A compilation of advice from experienced proofreaders and editors. Berkeley Heights (NJ) [accessed 2016 Nov 5]. http://www.lrcom.com/tips/proofreading_editing.htm

The Writing Center at UNC Chapel Hill. c2010–2014. Revising Drafts. Chapel Hill (NC): University of North Carolina; [accessed 2016 Nov 5]. http://writingcenter.unc.edu/handouts/revising-drafts/

The Writing Center at UNC Chapel Hill. c2010–2014. Editing and Proofreading. Chapel Hill (NC): University of North Carolina; [accessed 2016 Nov 11]. http://writingcenter.unc.edu/handouts/editing-and-proofreading/

The City University of New York School of Professional Studies Writing Fellows. c2016. CUNY Academic Commons. [accessed 2016 Nov 11]. https://bacwritingfellows.commons.gc.cuny.edu/proofreading/

Scientific style and format

Council of Science Editors, Style Manual Committee. 2014. Scientific style and format: The CSE manual for authors, editors, and publishers. 8th ed. Reston (VA): The Council of Science Editors. 722 p.

Peterson SM. 1999. Council of Biology Editors guidelines No. 2: Editing science graphs. Reston (VA): Council of Biology Editors. 34 p.

Peterson SM, Eastwood S. 1999. Council of Biology Editors guidelines No. 1: Posters and poster sessions. Reston (VA): Council of Biology Editors. 15 p.

Statistics

Moore DS, Notz WI, Fligner MA. 2011. *The Basic Practice of Statistics*, 6th ed. New York: W.H. Freeman and Company. 745 p.

Writing guides

Bullock R, Brody M, Weinberg F. 2014. *The Little Seagull Handbook*. New York: W.W. Norton & Co., Inc. 406 p. This manual has a section on Council of Science Editors style.

Hacker D, Sommers N. 2016. A Pocket Style Manual, 7th ed. Boston: Bedford/St. Martin's. 320 p. This manual has a section on Council of Science Editors style and a section on evaluating sources. This writing guide is my favorite.

Lunsford AA. 2016. The Everyday Writer, 6th ed. Boston: Bedford/St. Martin's. 656 p.

Lunsford AA. 2013. *Easy Writer*, 5th ed. Boston: Bedford/St. Martin's. 374 p.

Lunsford AA. 2015. *The New St. Martin's Handbook*, 8th ed. Boston: Bedford/St. Martin's. 912 p.

Index

A

Abbreviations
  for comments on lab reports, 150*t*–154*t*
  searches using, 17
  standard, 130*t*–131*t*
  use of, 129
Absorbance, definition of, 115
Absorbency, definition of, 115
Abstracts
  content checklist, 105*t*
  revision checklist, 138
  sample lab report, 142
  scientific paper, 33
  writing of, 87–88
Accuracy, technical, 109
Acknowledgements section, 34, 171
Active reading, 36
Active voice, 55, 87, 110–111
Adjectives
  coordinate, 126
  cumulative, 127
Adverbs, conjunctive, 126
Advertising, effective, 87
Affect, definition of, 115–116
AGRICOLA, 10*t*
Alphabetical order, 91, 94
Ambiguous usage, 113–114
"And" operator, 18–19
Animation, 180, 262, 263
Annual Reviews, 12, 23
APA (American Psychological Association), 95
Apple iCloud, 195*t*
Arial font, 165
Article databases, 10
Articles
  in books, citation of, 92*t*
  full-text, 23–24
  HTML option, 24
  reviews of, 31–32
Audiences
  active roles for, 180–181
  common knowledge of, 44
  "data dives" approach, 181*f*
  eye contact with, 182
  identification of, 53–54
  interaction with, 183
  lapses in attention, 180–181
  for poster sessions, 163
  questions from, 184
  slides friendly to, 176*f*, 177*f*
  slides geared to, 175–180
  speaker rapport with, 173
Authors
  listing on scientific papers, 33
  number of, 91
  revision checklist, 138
AutoCorrect, 193–197, 207, 208
Averages (means), 7
  calculation of, 251*f*
  error bars about, 251–253

B

Background information, 36
Backing up
  automated, 193
  methods for, 51–52
  options for, 194*t*–195*t*
  reliable, 193
Bar graphs, 70–72, 75*f*
  bar placement, 71
  clustered bars, 72*f*
  in Excel, 239–245
  horizontal bars, 71*f*
  Mac, 239–242
  purpose of, 61*t*

types of, 240*f*
Biological Abstracts, 10*t*, 12*t*
Biological Science (ProQuest), 10*t*
Biological sciences, information in, 10t–11*t*
BioOne, 10*t*t
Blackboard, 44, 184
Blank pages, errant, 205
BLAST database homepage, 98*f*
Body, oral presentations, 173–174
Books, citation of, 92*t*
Box, 195*t*

C

Calculations
of averages (means), 251*f*
rechecking, 105
recording of, 6
Capitalization, 63, 67
Captions. *See also* Legends
figure, 67
table, 63
Cardinal numbers, 129
Categorical data, 7
Categorical variables, 61, 70–72
Cell values, copying of, 220
Chart templates, 236–237
Chronological order, 106
Citation-Name system (C-N), 35, 90, 94, 97
Citation-Sequence (C-S) system, 92*t*, 93*t*
database citations, 98
description of, 28, 29*f*, 34
endnotes in, 197–198
homepage citations, 100
journal article citations, 97
use of, 90, 92–94
Citations
of articles in books, 92*t*
of books, 92*t*
of databases, 98–99
of discussion lists, 101
downloading, 25
of email information, 101
of homepages, 99–101, 100*f*
of Internet sources, 95–101
of journal articles, 96–97
management of, 24–30
of personal communications, 95
on posters, 171
of published sources, 58–59
of sources, 58–59, 86
tenses used on, 112
unpublished laboratory exercises, 94–95
Classrooms, note taking, 40
Clichés, use of, 120
Cloud storage, 195*t*
Clustered column charts, 245–246, 245*f*
College, success in, 42–43
Colons, use of, 127
Commas, purpose of, 125–127
Comments
abbreviations used for, 150*t*–154*t*
feedback using, 199–201
Common knowledge, 44
Communications, types of, 31–32
Comprehension, reading for, 36
Computers, understanding of, 51–53
Concentrations, solution, 57–58
Concepts
in the introduction, 85
mapping, 41–42
subdivision of, 16–17
Conclusion sections, on posters, 171
Connections, description of, 81–87
Control groups, 3, 4
Correlation values, 66
Council of Science Editors (CSE), 34–35
reference styles, 27
on use of color, 70
Counselling Services, University of Victoria, 35, 39
Credibility, reputation and, 47

D

Dashes, use of, 128–129
Data
comparisons, 77–78
erroneous, 7
multiple sets of, 69–70
organization of, 73–75
recording of, 5–6
singular of, 117
sorting of, 221
summarization of, 59
trustworthy, 7
Data analysis, 7
error bars and, 253
of raw data, 59
Databases
advanced search options, 23
article collections, 10–11

in the biological sciences, 10*t*–11*t*
citations of, 98–99
comparison of, 13–15, 13*t*
evaluating search results, 20*f*
for scientific information, 12–14
search strategies, 15–19
Dates, recording of, 6
Decimal numbers, 132
decimal places, 220–221
Dependent variables, 66
definition of, 4
Mac, Excel graph, 225*f*
Details, level of, 55–59
Dictionaries, use of, 114–115
Discussion lists, citation of, 101
Discussion section
chronology of, 83
content checklist, 105*t*
formulating questions for, 38
in need of revision, 160–161
order of, 107
on posters, 171
revision checklist, 139
sample lab report, 147–148
start of, 81
verb tenses in, 84
writing of, 81–84
Dissertations, 31–32
Documents
backing up, 193
formatting of, 201–204
naming, 192
proofreading of, 204–205
saving of, 192–193
DOI (digital object identifiers), 95, 96*f*
Drafts, oral presentation, 175
Drawings, recording of, 6
Dropbox, 195*t*

E

Editing
revision and, 104–121
time needed for, 49, 50*t*
Emails
backups using, 195*t*
citation of, 101
Empty phrases, 113, 114*t*–115*t*
End references, 89
Citation-Sequence system, 93*t*
documentation of, 91
Name-Year system, 93*f*
order of, 94
Endnotes, Citation-Sequence system, 197–198
Equations, in text, 80, 198–199
Error bars, 67
in Excel, 251–253
vertical, 252*f*
Errors, sources of, 8
Evaluation Forms, 185–187
on posters, 172
Examples, oral presentations, 183
Excel, Microsoft
adding chart elements, 227, 241, 243, 246–247, 248
adding data to existing graphs, 237–238
adding lines to existing graphs, 237–238
adding worksheets to file, 224
adjusting bar width, 242, 244
bar graphs, 239–245
changing bar color, 242, 244
changing legend titles, 238
chart borders, 230
chart templates, 236–237
choosing lines for graphs, 229–230
clustered column charts, 245–246, 245*f*
entering data, 243, 248
entering spreadsheet data, 239, 246
error bars, 249–253
formatting cells, 221
formatting graph axes, 228, 241–242, 244
formatting graph markers, 228–229
formatting lines for graphs, 230
formulas in, 214–220
graph data sets, 225–226
graphs in, 51
importing graphs into PowerPoint, 263
importing into Word documents, 242, 245, 247, 248–249
line graphs, 64
making graphs in, 213–253
page layout tab, 223
pie graphs, 246–247, 247*f*
plots on, 70
plotting data, 227
plotting graphs, 224–236
removing chart borders, 242, 244, 247, 248
saving chart templates, 242, 244, 247, 248

selecting data, 226–227, 239, 243, 246, 248
sorting data, 221
tables in, 221–223
views tab, 223
worksheet command area, 214*f*
XY graphs, 225*f*, 226*f*
XY graphs, multiple lines on, 239

Experiments
data recording, 5–6
definition of, 3
design of, 3–5
explanation of results, 7–8
field, 56
purpose of, 6, 53
treatments, 66
variables used in, 4–5

F

Feedback
from instructors, 135–137
oral presentations, 185–187
from peer reviewers, 135
on revisions, 133–137
using comments, 199–201

Feedback Studio, 137
Field experiments, 56
Figures
captions for, 67, 205
data summarized in, 59
formulating questions, 37
in lab reports, 52*t*
preparation checklist, 72
titles for, 67–70

Files
backing up, 193
naming, 192

Font sizes
choice of, 52*t*
for posters, 165
on slides, 179

Fonts
bold, 165
poster, 165

Format painter, 201
Formats, revision of, 132–133
Formulas
copying using the fill handle, 220
in Excel, 214–220, 217*f*
singular of, 117
writing for Windows, 218–219
writing in Mac, 215–216

Framing, posters, 164
Future work, 84

G

Gel images, 85
Gender-neutral language, 120
General-to-specific order, 107
Google, 36
Google Docs, 26
Google Drive, 51, 195*t*
Google Scholar
abstracts on, 22*f*
access to, 11
description of, 10*t*
full-text articles on, 23–24
resource type, 12*t*
results pages of, 21*f*
scope of coverage, 14
stemming technology, 19

GradeMark, 136–137, 137*f*
Grammar
checking, 204
definition of, 121

Grant proposals, 31–32
Graphics for posters, 164
Graphs
formulating questions, 37
Mac
data sets, 225–226
entering data, 224
pattern identification and, 7
plotting in Excel, 224–236
for PowerPoint posters, 166–168
recording of, 6
simplicity of, 177–179
types of, 60*t*
use of, 61

Group presentations, 184

H

Handouts
fonts for, 179
for presentations, 186*t*

Hanging indents, 202–203
Hard copy, printing of, 205
Hard drives, external, 194*t*
Headings in lab reports, 52*t*
Histograms, purpose of, 60*t*
Homepages, citation of, 99–101, 100*f*
Human errors, 82
Humor, in presentations, 183
Hypotheses
development of, 2–3
results and, 82

revision of, 8
singular of, 117
support for, 3
testing of, 3–5

I

Images
formulating questions, 37
for PowerPoint posters, 166–168
recording of, 6
IMRD format, 33
In-text references, 89
Citation-Sequence (C-S) system, 93*t*
Name-Year (N-Y) system, 93*f*
Independent variables, 66
definition of, 4
in graphs, 225*f*
hypothesis testing and, 3
treatment of, 5
Information
acknowledgement of, 44
background, 36
in the biological sciences, 10*t*–11*t*
hypotheses and, 3
lack of, 56–57
in oral presentations, 173
on posters, 164
too much, 57
Informational talks, 174–175
Instructional talks, 174–175
Instructions to Authors, 51
laboratory reports, 52*t*
Instructors, feedback from, 135–137
Interlibrary loans, 24
Internal servers, backups using, 194*t*–195*t*
Internet sources, citation of, 95–101
Introduction section
content checklist, 105*t*
details in, 86
formulating questions, 36–37
in need of revision, 155
to oral presentations, 173–174
order of, 106–107
organization of, 85
parts of, 33
on posters, 170
revision checklist, 138
sample lab report, 142–143
scientific paper, 33
voice in the, 87
writing of, 84–87
Introductory phrases, 79–80
Investigators, recording of, 6
Italicization
automatic, 196–197
of scientific names, 196–197

J

Jargon, definition of, 53, 120
Journal articles
citation of, 92*t*, 96–97
figure formatting for, 178*f*
oral presentations compared with, 174*t*
reading of, 35–39
JSTOR, 11*t*
Justification, in lab reports, 52*t*

K

Keywords, 36
database searches, 19
operators and, 18–19
Knowledge
common, 44
sharing, 39

L

Lab Report Template, 51
Lab reports
checklist, 138–140
structure of, 104–105
Labels, fonts for, 165
Laboratory exercises, unpublished, 94–95
Laboratory reports
abbreviations used for comments, 150*t*–154*t*
audience for, 53–54
data transfer into, 59
errors made in, 149–154
formatting, 51, 52*t*
getting started, 53
instructions to authors, 52*t*
in need of revision, 154–161
organization of, 53
preparation of, 49–101
sample, 141–161
timetable for, 49–53
timetable for preparation of, 50*t*
use of color in, 70
writing style, 54
writing titles for, 88–89
Learning, meaningful, 42–43
Legends. *See also* Captions
changing titles, 238

Letters of recommendations, 31–32
Libraries, topic clarification and, 15
Library of Congress catalog, 15
Library research skills, 9
Line breaks, 202, 203
Line (XY) graphs, 64–70
  data summarized in, 76*f*
  formats, 64–65*f*
  purpose of, 60*t*
  standard curves, 67
  straight line, 67
  as tools, 78–79
  use of, 7
Line indents
  in Windows, 202–203
Line spacing, Windows, 203
List of Authors, of scientific papers, 33
Literature searches, 9–30

M

Mac. *See also* Excel, Microsoft; PowerPoint, Microsoft; Word, Microsoft
  adding comments, 199–200
  anchoring images in tables, 209
  applying functions, 216–218
  automatic italization, 197
  creation of tables, 208
  deleting comments, 199–200
  endnotes in Citation-Sequence system, 198
  expressions with sub/superscripts, 196
  formatting paragraphs, 202
  formula tab, 216*f*
  grammar checking, 204
  hanging indents, 202
  inserting equations, 198–199
  inserting spreadsheet rows, 222
  line breaks, 202
  navigating in tables, 209
  page breaks, 202
  page numbers, 203
  paint bucket symbol, 228
  plotting graphs in Excel, 224–231
  repeating header rows in tables, 209
  setting margins, 201
  spell checking, 204
  subscript/superscript formatting, 206
  symbols in text, 206–207
  text within tables, 209
  tracking changes, 200
  writing formulas in, 215–218
Maps, 85
Margins
  choice of, 52*t*
  setting, 201
Markup, tracking changes, 199
Materials, listing of, 57
Materials and Methods section
  content checklist, 105*t*
  formulating questions, 37
  levels of detail in, 55–59
  in need of revision, 155
  on posters, 170
  revision checklist, 138–139
  sample lab report, 143–145
  scientific paper, 33–34
  verb tenses, 54, 55
  voice in, 55
Math symbols, inserting equations, 198–199
Means (averages), 7
  calculation of, 251*f*
  error bars about, 251–253
Measurement, units of, 6
Medical Subject Headings. *See* MeSH headings
Mega, 195*t*
MeSH Database, 17
MeSH headings, 14, 16–17, 18*f*
Microsoft OneDrive, 195*t*
Mind mapping, 41–42
Model papers, 47
Moodle, 184
Mozy, 195*t*

N

Name-Year (N-Y) system
  database citations, 98
  description of, 29–30, 30*f*, 35, 90–92, 91*t*, 92*t*, 93*f*
  homepage citations, 99–100, 101
  journal article citations, 97
National Center for Biotechnology Information, 11t, 98*f*
Negative controls, 4
"Not" operator, 18–19
Notes
  presentation view, 185
  taking of, 40
Null hypotheses, 5
Numbers
  spelled out, 132
  use of, 129
Numerical data

entry errors, 82
summarized, 6–7

O

Observance
definition of, 115
use of, 118
Observational studies, 65
Opening sentences, 86
Operators
used in Excel, 215*t*
used in searches, 18–19
"Or" operator, 18–19
Oral presentations, 173–187
body, 173–174
closings, 174
delivery of, 182–184, 267
ending of, 183
environment for, 174–175
feedback, 185–187
figure formatting for, 178*f*
group, 184
introduction, 173–174
journal articles compared with, 174*t*
organization of, 181–182
planning for, 174–175, 258
presentation style, 182–183
rehearsals, 181–182, 267–268
resources available for, 269
results shared through, 8
timing of, 181
tips, 267*t*
title slides, 260–261
use of examples in, 183
using PowerPoint, 255–271

P

Page breaks, 202, 203
Page numbers, 203–204
Pagination in lab reports, 52*t*
Paint bucket symbol, 228
Paintbrush, 201
Palatino font, 165
Paper, choice of, 52*t*
Paragraphs
formatting of, 202–203
structure of, 106–108
Paraphrasing, source texts, 45
Parentheses, use of, 128
Parenthetical expressions, 126
Passive voice, 55, 87, 110–111
Past tense, 54, 79–80, 111–112
PDF files, 24
sticky notes in, 136
Peer reviewers, discussion with, 135
Peer reviews
time needed for, 50*t*
tips, 134–135
Periods, use of, 128
Personal communications, citation of, 95
Persuasive talks, 174–175
Phenomena, 2
Photographs, 85
Phrases, empty, 113
Phylogenetic trees, 85
Pie graphs, 61*t*, 72–73, 73*f*, 246–249, 247*f*
Plagiarism, 44
checking for, 136
examples of, 46*t*
Plurals
frequently used, 118*t*
use of, 118
Pointers, 182
Positive controls, 4
Poster sessions, 8
Posters
appearance, 164
benefits of, 163
content, 169–171
display of, 166
evaluation forms, 172
final printing, 169
formats for, 164
layouts, 164
presentations, 163–172
proofreading, 168–169
samples, 172
title banners, 169–170
PowerPoint, Microsoft
adding slides, 261
adding text, 264
adding videos, 265–266
aligning objects, 168, 168*f*
aligning shapes, 265
animation on, 180, 262, 263
chart tools, 167
command area, 256*f*
drawing tools, 167
file handling, 257
grouping shapes, 264
importing Excel graphs, 263
laptop presentations, 269–270
merging shapes, 264
Notes pane, 272
for oral presentations, 175

pen tool, 270*f*
pictures, 167
poster preparation using, 166–169
preparing oral presentation, 255–271
presenter tools, 270–271
Presenter View, 187*f*, 267, 268*f*, 269, 270*f*
rehearsals, 267–268
reusing slides, 261–262, 262–263
shapes, 168
slide show mode, 269–271
slides in class, 40
speaker notes, 272
text formatting, 266
theme selection, 259
title slides, 260–261
using lines, 263–264
Web server presentations, 270
working with shapes, 263–267
Prereading method, 35
Present tense, 54, 111–112
Presentations. *See* Oral presentations
Primary references, 9
Printing hard copy, 205
Procedures
in reports, 58
sequential order of, 106
Pronouns, ambiguous use of, 113–114
Proofreading
checklist for, 133*t*
of documents, 204–205
editing and, 104
format checks, 132–133
keeping a log of issues, 135
marks, 149–150, 150*t*
posters, 168–169
steps to follow, 121–133
time needed for, 49, 50*t*
ProQuest, 24–30, 26*f*
Publication of results, 8
PubMed
abstracts on, 22*f*
access to, 11
comparison of, 13*t*
description of, 11*t*
full-text articles on, 23–24
homepages, 17*f*
results pages of, 20, 21*f*
scope of coverage, 14
Similar Articles option, 23
Punctuation, purpose of, 125–129

Q

Quantitative variables, 61, 64
Questions
answers to, 2
from the audience, 184
formulation of, 36–38, 40
unanswered, 85

R

Reading
reviewing information, 38–39
selective, 38, 40–41
textbook, 39–42
Recitation, reviewing information, 38, 41
Redundancy
avoiding, 108–109
examples of, 112*t*
wordiness and, 112–113
Reference section
in need of revision, 161
revision checklist, 140
sample lab report, 148–149
References
content checklist, 105*t*
insertion in end references, 27–30
insertion in-text, 27–30
in lab reports, 52*t*
management of, 24–30
order of, 34–35
primary, 9
scientific paper formats, 34
secondary, 9
RefWorks
Citation Manager, 25–27
Download Citations box, 25, 25*f*
for iPad, 27
login page, 24
ProQuest, 24–30, 26*f*
tutorials, 24
Regression lines, 61*t*
Rehearsals, presentation, 181–182
Relationships
illustration of, 76
strength of, 66
Replicates
averaging of, 59
hypothesis testing and, 3
statistical validity and, 5
Reputation, credibility and, 47
Research, goals of, 16
Research proposals, 31–32

Results
explanation of, 7–8
interpretation of, 81–84
Results section, 34
content checklist, 105*t*
format of, 62*f*, 68*f*, 69*f*
formulating questions, 37–38
in need of revision, 156–159
on posters, 170–171
revision checklist, 139
sample lab report, 145–147
verb tenses, 54
writing of, 59–80, 75–79
Review articles, use of, 23
Reviews, 31–32
Revisions, 103–140
audience considerations in, 104
checklist, 140
definition of, 103
editing, 104–121
feedback on, 133–137
preparation for, 103–104
of sentences, 122–124
time needed for, 50*t*
Ribbon interfaces, 189, 213
Rows, spreadsheet, 222

S

SafeAssign™, 44
Sample lab report in need of revision, 154–161
Sample sizes, 82
Scatterplots, 64–70, 250*f*
form of, 66
numerical data on, 66
purpose of, 60*t*, 61*t*
use of, 7
ScienceDirect, 11*t*
Scientific information
databases for, 12–14
search engines for, 12–14
Scientific method, steps in, 1–8
Scientific names, 196–197
Scientific notation, 132
Scientific papers
formats of, 32–34
learning to write, 46–47
models for, 47
preparation of, 49–101
reading of, 31–47
structure of, 104–105
timetable for, 49–53
writing of, 31–47
writing titles for, 88–89
Scientific terminology
definition of, 120
spell checkers and, 124–125
Scientific writing, hallmarks of, 32
Scopus, 11*t*, 13*t*, 15
SD/microSD cards, 194*t*
Search engines
advances search options, 23
in the biological sciences, 10*t*–11*t*
evaluating results, 19
evaluating search results, 20*f*
operators used in, 18–19
for scientific information, 12–14
term choices, 17
Searches, evaluating results, 19
Secondary references, 9
Section headings, 204–205
Semicolon, use of, 127
Sentences
complete, 122
complexity in, 109–110
length of, 109–110
meaningful, 108–112
order of, 106–107
run-on, 122–124
subject of, 110
verb order, 110
Sequential ordering, 106
Serifs, 165
Signal phrases, 107–108
Signal words, 107–108
Sketches in lab reports, 52*t*
Slang, use of, 120
Slide decks
description of, 174
note pages, 185*f*
as references, 184–185
sharing, 186*t*
Slide shows, 255. *See also* Oral presentations
Slides
addition of, 260–261
animation features, 180, 262, 263
audience and, 175–180
design of, 176, 258–267
layouts, 260–261
slide size, 260
theme selection, 258–259
views, 260
sizes, 166, 167*f*

text on, 179
titles for, 177–178, 178*f*
white space on, 179
Software, understanding, 51–53
Solutions
concentrations of, 57–58
preparation of, 57–58
Sources
alphabetical order, 91
citation of, 58–59, 86
documentation of, 89–95
paraphrasing of, 45
reference sections, 34
Spacing in lab reports, 52*t*
Speakers
attire, 182
notecards, 182
posture of, 182
slides friendly to, 176*f*, 177*f*
visuals, 182
Specific-to-general order, 107
Spell checkers, 124–125, 204
Spreadsheets. *See also* Excel, Microsoft
format cells, 221
formatting of, 220–224
inserting rows, 222
sorting data, 221
SQ3R method, 39–40
Standard curves, 67
Standard deviation, 251*f*
Standard error, 251*f*
Statistical inferences, 6
Strain, definition of, 119
Strand, definition of, 119
Student lab reports
good sample, 141–149
in need of revision, 154–161
Study groups, 43
Studying, 43
Subheadings in lab reports, 52*t*
Subjects, verb agreement with, 121–122
Subscripts
expressions with, 196
formatting of, 205–206
SugarSync, 195*t*
Superscripts, 196, 205–206
Symbols
AutoCorrect for, 207, 208
choice of, 52*t*
in text, 206–208

T

Tables
anchoring images in, 209
captions for, 63, 205
creation of, 208–211, 210
data summarized in, 59, 76*f*
definition of, 60
in Excel worksheets, 221–223
formatting text within, 209
in lab reports, 52*t*
navigating within, 209
preparation checklist, 64
repeating header rows, 210
titles, 63–64
use of, 7, 60–64
Teaching and Learning Center, Bucknell University, 35, 39
Technical accuracy, 109
Templates
for charts in Excel, 236–237
in PowerPoint, 175
Tenses, use of, 111–112
Terms, choice of, 17
Text
for PowerPoint posters, 166–168
on slides, 179
wrapping, 220
Textbooks
background information on, 36
formulation of questions, 40
instructions for using, 43
reading strategies for, 39–42
reviewing information, 41
selective reading of, 40–41
surveying content of, 39–40
Than, use of, 119
That, use of, 119
Theses, 31–32
Times font, 165
Title pages in lab reports, 52*t*
Titles
capitalization of, 63
content checklist, 105*t*
fonts, 165
revision checklist, 138
of scientific papers, 33
for slides, 177–178, 178*f*
table, 63–64
writing of, 88–89
Topics
clarification of, 15
ordering of, 106

subdivision of, 16–17
Tracking changes, comments and, 199–201
Transitional expressions, 126
Transitions, use of, 107–108
Treatment groups
hypothesis testing and, 3
responses and, 66–67
set up of, 4
Truncation symbols, 19
Turnitin®, 44, 136, 137
Typefaces, choice of, 52*t*

U

URL addresses
citing Internet sources, 95–101, 96*f*
ending of, 16*t*
in-text, 96
website reliability and, 15
USB flash drives, 51, 194*t*

V

Variability of data, 7
Variables
categorical, 61, 70–72
definition of, 4
dependent, 4, 66, 225*f*
formulating questions, 37
independent, 3, 4, 5, 66, 225*f*
quantitative, 61, 64
Verbs
order in sentences, 110
subject agreement with, 121–122
tenses, 54, 84, 86, 111–112
Videos in PowerPoint, 265–266
Visuals
descriptions of, 59–60
editing of, 104
memorable, 120–121
for oral presentations, 173
positioning of, 121
for poster sessions, 163
posters, 164
purpose of, 121
text replaced with, 177
Voice in the introduction, 87

W

Web of Science
abstracts on, 22*f*
Cited References section, 23
comparison of, 13*t*
description of, 11*t*
full-text articles on, 23–24
results pages of, 20, 21*f*
scope of coverage, 14
Times Cited section, 23
View Related Records section, 23
WebMD, 15
Websites
reliability of, 15
sponsorship of, 15, 16*t*
White space on slides, 179
Wikipedia, 15, 36
Windows. *See also* Excel, Microsoft; PowerPoint, Microsoft; Word, Microsoft
accessing files/folders, 191–192
adding chart elements, 232–233
adding comments, 200
anchoring images in tables, 211
applying functions, 218–219
AutoCorrect on, 195
automatic italization, 197
choosing lines for graphs, 234–235
deleting comments, 200
Document Inspector, 200
endnotes in Citation-Sequence system, 198
entering data for graphs, 231–232
expressions with sub/superscripts, 196
file organization, 191
formatting graph axes, 233
formatting lines for graphs, 235
formatting markers, 233–234
formatting paragraphs, 202–203
formatting text in tables, 210
grammar checking, 204
hanging indents in, 202–203
importing graph into Word, 235–236
inserting equations, 199
inserting spreadsheet rows, 222
line breaks in, 203
line indents in, 202–203
lines in a table, 210–211
navigating in tables, 211
page breaks in, 203
page numbers, 203–204
paint bucket symbol, 233–234
paragraph formatting, 202–203
removing chart borders, 235
repeating header rows in tables, 211
selecting data for graphs, 232
setting margins, 201

spell checking, 204
subscript/superscript formatting, 206
symbols in text, 207–208
tracking changes, 200–201
writing formulas in, 218–219

Word, Microsoft
accessing files/folders, 191–192
citation management, 26, 27*f*
command area, 190*f*
Equation Editor, 80
file organization, 190–191
finding commands in, 189
formatting commands, 51
good housekeeping in, 190–193
importing Excel graphs into, 231
making comments, 134–135
organizing files, 190–191
reviewing electronic files, 134–135
saving documents, 192–193
tables in, 64
tracking changes, 134–135, 136
word processing in, 51, 189–211

Word processing. *See also* Word, Microsoft
accessing files/folders, 191–192
file organization, 190–191
formatting documents, 201
naming files, 192

Words, choice of, 112–120
World Wide Web, 16*t*
Wrap text, spreadsheet cells, 220
Write-N-Cite, 25–27
Writing styles, laboratory reports, 54

X

XY graphs, 225*f*, 226*f*, 243–245
symbol hierarchy, 229*t*

Y

YouTube, 36